Inclusive by Design

Inclusive by Design

Crafting Products and Services
for a More Equitable World

Mathieu Aguesse

WILEY

Published by John Wiley & Sons, Inc., Hoboken, New Jersey.
Published simultaneously in Canada and the United Kingdom.

ISBNs: 9781394310050 (Paperback), 9781394310074 (ePDF), 9781394310067 (ePub)

For general information on our other products and services or for technical support, please contact our Customer Care Department within the United States at (800) 762-2974, outside the United States at (317) 572-3993 or fax (317) 572-4002. For product technical support, you can find answers to frequently asked questions or reach us via live chat at https://support.wiley.com.

If you believe you've found a mistake in this book, please bring it to our attention by emailing our reader support team at wileysupport@wiley.com with the subject line "Possible Book Errata Submission."

Wiley also publishes its books in a variety of electronic formats. Some content that appears in print may not be available in electronic formats. For more information about Wiley products, visit our web site at www.wiley.com.

Library of Congress Control Number: 2025932718

Cover image: Created by Jan Veicht
Cover design: Wiley

SKY10100391_031725

About the Author

Mathieu Aguesse is an entrepreneur, educator, designer, and strategic consultant dedicated to fostering Inclusive and Equitable Design. With a multicultural background spanning Africa, Europe, and the United States, he has worked at the intersection of technology, design, and social impact, bringing together diverse stakeholders—from large corporations and startups to universities and policymakers—to create systemic, sustainable change.

Over the years, Mathieu has worked with hundreds of companies and leaders, bringing his expertise in innovation and design to support, consult, and challenge them in their strategic endeavors. He has drawn from, refined, and expanded upon a range of frameworks and techniques, including Design Thinking, Sustainable Design, Design Fiction, Future Modeling, and, of course, Inclusive and Equitable Design—adapting existing methodologies while developing his own. His work has influenced global leaders across industries—automotive, retail, food, real estate, finance, and beyond—helping them reimagine the future, anticipate emerging challenges, and create products and services that drive a more sustainable and equitable society.

Since 2018, he has played a pivotal role in the San Francisco innovation ecosystem, founding and managing an innovation hub called Schoolab that has connected European institutions, Fortune 500 companies, cutting-edge startups, and students. His dedication to education is evident through his long-standing role at UC Berkeley since 2019, where he founded two programs—Deplastify the Planet and the Equitable Design Lab—and taught up to four classes, as well at MIT, where he taught Design classes within the D-Lab.

Mathieu's passion for diversity, equity, and inclusion (DEI) extends beyond theory—he has worked with governments, businesses, and NGOs to embed DEI principles into strategic decision-making, policy design, and product development.

With this book, Mathieu aims to bridge the gap between theory and action, offering tools, frameworks, and methodologies that empower organizations and individuals to design for inclusion, with inclusion, and by inclusion. Through his work, he continues to challenge dominant narratives, pushing for a world where design is not just an afterthought but a driving force for this essential and society needed equity and systemic transformation.

Engineer and designer by training, changemaker by passion. CEO and teacher during the day, father of three at night.

About the Technical Editor

Barry Katz is Emeritus Professor of Industrial and Interaction Design at the California College of the Arts in San Francisco and formerly Adjunct Professor in the Design Group, Department of Mechanical Engineering, at Stanford University and Lecturer at the Jacobs Institute for Design Innovation at U.C. Berkeley. As an IDEO Fellow, Barry worked for more than 20 years with the world's leading design and innovation consultancy, and he continues to consult with governments, companies, and academic institutions worldwide. His writings on design as a strategy of innovation have appeared in many academic, professional, and popular journals.

Acknowledgments

This book is the result of an unexpected, high-speed, winding journey shaped by some brilliant minds, unwavering support, and deep generosity of many people along the way. I am profoundly grateful to all those who contributed to making this work a reality, sometimes against the odds!

First, my deepest thanks to Ibrahim Balde, my former student and close collaborator and now a lifelong friend. Together, we built this program and framework, and without his insights, energy, and belief in this vision, none of this would have come to life. His journey has been a constant source of inspiration, pushing me to refine, challenge, and expand what Inclusive Design could truly mean.

To Jan Veicht, Cian Mitsunaga, Kelly Chou, Patrick Consorti, and the entire Schoolab San Francisco team—thank you for your support and creativity and for bringing your unique perspectives and skills to this project. Your belief in this work and your dedication to making Inclusive Design a tangible reality has been invaluable. Your encouragement and critical thinking helped shape not only this book but also the broader movement we've built together.

To my students, partners, and collaborators—you are the heartbeat of this work. Without your trust, engagement, and commitment, this journey would be meaningless. You have invested your time, resources, and skills to bring Inclusive Design to the world, proving that change happens when people come together with purpose and action.

To my family—my incredible wife, who has been my steadfast supporter, believing in me through every challenge and ambition, and my three children, whose very existence pushes me to dedicate my work to meaningful, long-term change. My hope is that one day, when they are grown, they will understand how much they influenced me to focus on causes that truly matter.

A special thank you to Barry Katz, Marco Lindsey, and the Wiley team, whose belief in this book gave me the opportunity to share these ideas with the world.

Finally, my gratitude extends to Jean Louis Soubret and Clark Kellogg, whose wisdom and guidance supported me when I needed it most. Your mentorship and insights have left an indelible mark on my thinking and approach.

To everyone who has contributed, supported, or inspired this work—thank you. This book is not just mine; it belongs to the many voices, hands, and minds that shaped it.

Contents at a Glance

Contents

Introduction

Growing up as a left-handed person, I was exposed early on to a world where many objects and systems were simply not designed for me. One of my first memorable encounters with this reality came when I was about five years old and in kindergarten. My teacher, determined to teach me how to cut with scissors, insisted that I use my right hand—completely ignoring my natural left-handedness. Although I wouldn't have called it this at the time, it was my first direct experience with what I now recognize as a non-inclusive design.

Yet I never felt like a victim. In fact, I remember an instance when I was tinkering with my grandfather and we needed to screw two boards together in a tight cupboard space. The awkward angle made it nearly impossible for a right-handed person, and he turned to me, saying, "This is a task for you—only a left-handed person can do it!" In that moment, what could have been a limitation became a unique advantage, showing me how context could shift perception.

At the same time, my childhood in South Africa during the mid-1990s gave me a front-row seat to one of the most profound movements for equality in modern history. It was 1995, and Nelson Mandela had just taken office as the country's president. The dismantling of apartheid—a deeply entrenched and systematic form of inequality—was underway. Yet as transformative as those changes were, decades later, South Africa remains riddled with inequalities. The visible infrastructure of inequality may have been dismantled, but deeper, systemic layers continue to perpetuate disparities.

One of those layers is Design: the way products, services, and systems are created. Design has a profound ability to either support inclusion and equality or reinforce barriers and inequities. For me, this realization sparked a deep interest in how intentionality in design could be a force for social good.

The physical ergonomics of objects are often the first things that come to mind when thinking about inclusive design. From scissors that are impossible to use with the left hand to stairs-only pathways that exclude wheelchair users and parents with strollers, poor physical design can create unnecessary barriers. But inclusion extends far beyond ergonomics.

Take accessibility of services, for instance. The location and hours of public services can alienate individuals who work long shifts or live in areas without reliable transportation. Similarly, digital services requiring fast and stable internet access often exclude those in underserved or rural areas. These barriers highlight how poor design decisions can restrict access to critical resources.

Education and literacy levels are another overlooked design factor. Consider financial services: someone raised in a privileged environment may find concepts like credit, interest, and savings intuitive. But for someone with little financial literacy, this jargon can be overwhelming, further distancing them from opportunities for economic stability.

These examples reveal how poor design can perpetuate exclusion and inequality. As the saying goes, "Design is the intention behind an outcome." Inclusive Design, at its core, is about intentionality: creating products, services, and experiences that break down barriers and empower those who face the most challenges. It's not just about functionality but about fostering equity and inclusion through thoughtful, empathetic creation.

We live in a very interesting time where we have access to more information than ever and the ability to draw better conclusions from history and leverage science to anticipate some of the threats ahead of us. Yet we are also experiencing a widespread rejection of democracy and progressive ideologies. Recent elections in the Western world demonstrate growing disapproval of elites, a sentiment likely fueled by unprecedented levels of inequality in wealth and power.

As of early 2024, in the United States, the wealth disparities in society are stark. The top 10% of households by wealth control an astounding 67% of total household wealth. Meanwhile, the bottom 50% of households collectively hold just 2.5% of total wealth. This disparity is not a recent phenomenon but part of a long-term trend; the wealth gap between America's richest and poorest families more than doubled from 1989 to 2016, highlighting the deep structural inequities at play.[1, 2]

Racial wealth disparities in the United States remain significant. According to a 2023 report by the Pew Research Center, in 2021, the typical White household had $201,700 more in wealth than the typical Hispanic household, up from a $168,800 gap in 2019. Similarly, data from the Federal Reserve Bank of St. Louis in 2024 indicates that Black families, on average, owned about 23 cents for every dollar of White family wealth, and Hispanic families owned about 19 cents per dollar. These figures reflect a persistent pattern of wealth concentration and systemic exclusion, with profound implications for social cohesion, democratic stability, and economic mobility.[3, 4]

In such an environment, it is unsurprising that frustrations are mounting, further driving divisions and skepticism toward established systems and institutions.

The risks posed by deep and persistent inequalities are immense and often underestimated in modern society. History provides us with stark warnings: major civilizational crises and eventual collapses often occurred when societies became profoundly stratified. In such scenarios, a small elite accumulated power and resources, and the majority of the population was left vulnerable. When these societies faced hardships such as prolonged droughts, economic stagnation, or other climate-related catastrophes, their resilience eroded, and the imbalance prevented effective collective responses.

Take the breakdown of Rwanda in the 1990s, for instance. As Jared Diamond discusses in *Collapse: How Societies Choose to Fail or Succeed* (2011, Penguin Publishing Group), extreme inequalities in land ownership and resource distribution, combined with rapid population growth, created unsustainable pressure on an already fragile system. The severe competition for land exacerbated deep-rooted ethnic tensions, ultimately leading to the 1994 genocide, one of the most devastating societal collapses in modern history. Similarly, the Roman Empire, despite its immense power, succumbed in part to growing economic disparity—a widening gap between the wealthy elite and the struggling lower classes—alongside environmental stress and external invasions. As the rich consolidated power and resources, the empire lost the social cohesion and adaptability necessary to withstand external shocks.

These examples illustrate a dangerous pattern: when inequalities deepen, societies lose their resilience, their ability to adapt to crises, and their sense of collective purpose. Without inclusive and equitable systems, civilizations become brittle—prone to internal strife and incapable of withstanding external pressures, ultimately leading to their downfall.

In today's world, these lessons are especially relevant. Climate change, resource depletion, and political instability are pressing global threats. If left unaddressed, growing disparities in wealth and power could undermine collective efforts to mitigate these challenges. A society where large portions of the population feel excluded or disenfranchised lacks the solidarity and trust needed to navigate crises. Inequality is not just an ethical issue; it is a structural weakness that threatens the very fabric of civilization.

The purpose of this book is to showcase how thoughtful design can act as a powerful tool to create a more inclusive and equitable society. By drawing on a wide range of projects I've been involved in over the years, I aim to demonstrate the tangible ways in which Design, when approached with intention, can break down barriers and empower underserved communities. Through my teaching roles at two of the world's most prestigious institutions—UC Berkeley and MIT—and collaborations with some of the largest global companies, I've developed a framework and philosophy tailored to inspire and guide leaders, decision-makers, and designers alike.

Since graduating from graduate school, I have been intensely focused on elevating the role of Design in the business world. My driving obsession has been to challenge the status quo, where designers are often undervalued and perceived as creators of visually appealing artifacts rather than contributors to strategic decision-making. Too often, designers are excluded from the processes where overarching strategies are defined, leaving them to execute decisions made by others. Over the past 12 years, I've worked tirelessly to prove this perspective wrong and to demonstrate that Design is not just a tool for execution but a vital strategic imperative for businesses and institutions.

Operating at the intersection of studio design, strategic consulting, and education, I have combined creative insights, business acumen, and pedagogical methods to shift perspectives. My goal has been to convince those in positions of power that Design deserves a central role in shaping strategy. By merging these approaches, I've been able to show that when Design is harnessed effectively, it becomes a transformative force that can drive innovation and inclusivity on a broad scale. This book is the culmination of that journey: a roadmap for those who seek to use Design as a catalyst for societal progress.

This book is not merely a collection of ideas but also a hands-on framework designed to reimagine how we approach Design, with a focus on distributivity, additionality, inclusion, and impact. My goal is to provide practical tools and inspiring stories that empower designers, leaders, and engaged citizens to contribute to a more inclusive and equitable society.

Throughout these pages, I share a wealth of experiences gathered from an extraordinary journey at the intersection of innovation, design, and corporate responsibility. Between 2020 and the book's publication, I had the privilege of collaborating with over 50 companies, supporting their efforts to integrate Inclusive and Equitable Design into their products and services. In most cases, my team and I were tasked with understanding the systemic barriers and inequalities embedded within their markets and offerings. From this foundation, we developed strategies, crafted solutions, and introduced meaningful adjustments to reverse these trends—transforming exclusive experiences into empowering ones. This multifaceted work required us to address functional, technical, and business dimensions, ensuring a holistic approach to change.

I was fortunate to partner with an impressive roster of organizations, including Fortune 500 companies, trailblazing startups, global leaders in fast moving consumer goods (FMCG) and food and beverage, and some of the most renowned tech companies worldwide. These collaborators demonstrated an admirable openness and commitment to driving positive change, even when the conversations were challenging. The success of these projects was only possible due to the exceptional talent, dedication, and collaborative spirit of my team and theirs, navigating complex and often sensitive discussions with grace and insight.

To respect the confidentiality and integrity of these partnerships, I have chosen not to name any of the companies involved. Instead, I leave it to them to share their commitments and achievements in their own words and on their own timelines. This decision reflects my belief in the importance of authenticity and accountability in driving long-term, meaningful change.

My work as an educator at UC Berkeley and my collaborations with major global corporations have given me a unique vantage point to observe the power of design in driving transformative change. I've witnessed how seemingly small shifts in design thinking can lead to profound impacts—not only in the functionality of products and services but also in enhancing the quality of life for countless individuals.

Being an award-winning teacher at the world's most prestigious public university carries immense responsibility. When I step into a lecture hall or present to a room of eager minds, I am deeply aware that I'm not just delivering content—I'm planting seeds of ideas that may shape future leaders and innovators. Similarly, when advising VP and C-suite executives of global companies, the stakes feel equally high. I strive to ensure that my strategic recommendations not only serve my clients' business interests but also contribute positively to society at large.

As the CEO of my company, I endeavor to lead with intentionality in every decision I make, from the strategies we adopt to how we engage as a team. I'm fortunate to work alongside an exceptionally talented group of colleagues whose insights and perspectives profoundly influence my work. Their contributions keep me grounded and constantly push me to improve.

Finally, as a father of three and an active member of my community, I take my personal responsibilities very seriously. Every day, I strive to raise my children with integrity, empathy, and curiosity, hoping they will grow into individuals who contribute meaningfully to the world. In parallel, I seek to foster a thriving and inclusive community, believing that the values we model in our daily lives ripple outward to shape a more equitable future. These roles—educator, leader, parent, and community member—are all interconnected, forming the foundation of my commitment to a life driven by purpose and impact.

I also draw a lot of resources from my personal background. Growing up in Africa with French parents and then living in the United States, I experienced first-hand what diversity of culture can be. Living in many different places in my life and traveling the world as few people have the chance to do gave me the opportunity to experience a very unique range of situations where human beings come together. From interacting with kids at school in a remote village in rural Benin to attending private parties in Silicon Valley surrounded by millionaires, diversity and equality are concepts with which I am very familiar from a local and a global standpoint.

Being colorblind and left-handed, I know what it means to live in a world that is not designed for me. Yet I'm extremely cognizant of the incredible privileges I have. My students know this because I introduce myself in the first class by telling those future leaders that I am probably the most privileged person they've ever met: white, male, highly educated, with a healthy family, well paid, in great health, and so on for a long time.

I had multiple motivations for writing this book: sharing about the work I've been doing with top leaders at UC Berkeley, MIT, and large companies; and making sure some of the incredible content and learnings we have developed in this academic, business, and human experience are shared more broadly than within the community of partners, students, and teammates.

Writing those lines, the prime audience I had in mind was leaders across the globe who are making decisions that influence products, services, and processes. Without knowing it, they hold important keys to the complex equation of inclusion and equality in our society. They are the silent majority of people who can contribute to reducing inequalities and empowering the communities in need. They have a superpower in their hands, but it takes two things for them to tame it.

One is to understand what Design is and how much Design can influence our society. Design decisions shape the billions of products, services and processes we all use on a daily basis. Understanding the role of Design, the process and the philosophy of those who practice Design, is essential to realize the power of decision-makers in big and small companies across the globe.

The second step is to understand how Design can be a great ally for a more inclusive and equitable world. On the flip side, being a non-intentional designer can lead to perpetuating barriers and exclusion.

In this process, designers and business partners act as essential stakeholders and contributors. They are also people I would love to touch with my book. Adjusting their practice and becoming ambassadors of Inclusive Design is key in the process of change, and I hope they will find in this book some bulletproof cases and business arguments to convince their peers and clients to make more conscious choices.

Last, students and the younger generation have always been close to my heart: an incredible crowd to learn from and also what I've always called the leaders of tomorrow. It is probably too late to change the products and services of yesterday, it is possible to influence the products of today, and it is the right time to improve the products of tomorrow, and this can only happen through this next generation.

This book is divided into three parts. The first explores the principles and philosophy behind Inclusive and Equitable Design. The second delves into actionable strategies and case studies from my work with students and organizations. Finally, the third envisions a future where Design becomes a driver of equality, shaping a better world for all. I wanted to structure the book as I structure a project I'm working on. We start with the intention, the mindset, setting clear goals and a philosophy for success. Then we move to action, entering with humility and developing our solutions using tools, methodologies and learning from others. Next we create the infrastructure so that people can embark on our journey, and we make sure we always stay on track with our goals and intentions.

Last, we reflect on the journey and ask ourselves what it means and how our solution could impact things in the future. We expand the horizon and challenge ourselves to go further the next time.

Each part of the book includes real-life use cases filled with details and important learnings. Each chapter is a stepping stone to the next one. You can read this book in one piece or jump straight to chapters that could be directly relevant to your current activity and contribute to your project.

My hope is that this book doesn't just inform you but inspires you to act. Whether you're a designer, a leader, or simply someone who believes in a fairer world, this book offers tools and ideas to help you create meaningful change.

Notes

[1] Franklin, A.W., 2012. Management of the problem, in Smith, S.M. (ed.) *The maltreatment of children*. MTP, 83–95. Available at: https://www.stlouisfed.org/community-development-research/the-state-of-us-wealth-inequality.

[2] Schaeffer, K., 2020. 6 facts about economic inequality in the U.S., Pew Research Center. Available at: https://www.pewresearch.org/short-reads/2020/02/07/6-facts-about-economic-inequality-in-the-u-s.

[3] Duke News, 2024. U.S. racial wealth gap is persistent and growing, new research finds. Duke University. Available at: https://news.duke.edu/stories/2024/06/10/u-s-racial-wealth-gap-is-persistent-and-growing-new-research-finds.

[4] Menasce Horowitz, J., Igielnik, R., & Kochhar, R., 2020. Trends in income and wealth inequality. Pew Research Center. Available at: https://www.pewresearch.org/social-trends/2020/01/09/trends-in-income-and-wealth-inequality.

The Fundamentals of Inclusive Design

As I sat in my office, staring at the screen, I could feel the tension rising within me. The world was still reeling from the pandemic, and here I was, about to pitch an idea that had the potential to reshape the way we think about design. My fingers drummed nervously on the desk as I waited for the Zoom call to start. The once-familiar environment of my office felt strangely alien, the silence almost deafening as I rehearsed my points in my head. The uncertainty of the times only added to my anxiety.

I was about to propose a new course, "Equitable Design," to the senior vice-dean of the UC Berkeley Haas Business School. She was a woman of color renowned for her sharp intellect and unwavering efficiency. She commanded respect across the campus, and I knew this pitch had to be perfect. The stakes felt impossibly high. As I adjusted my camera and checked my microphone for what felt like the hundredth time, I couldn't help but reflect on how far I'd come and how much this moment meant.

The events of the past year had been transformative for many, including myself. The pandemic had forced us all into a new way of living and working, highlighting the deep-seated inequalities in our society. The tragic death of George Floyd had ignited a global movement, pushing issues of racism and social justice to the forefront of our collective consciousness. It was a time of reckoning, a time to reassess our values and actions.

This backdrop made my proposal even more significant. The idea of "Equitable Design" was not just a course; it was a call to action. It was a response to the pressing need for change in how we approach design and inclusivity. I knew that the dean, with her profound understanding and commitment to these issues, would grasp the importance of what I was trying to achieve. Yet the pressure to convey this effectively was immense.

In this chapter, we will delve into the origins of Inclusive Design through the lens of my personal journey as a teacher, entrepreneur, and seasoned strategic consultant. Over the years, I have supported dozens of Fortune 500 companies in their efforts to become more inclusive, and this experience has shaped my perspective on the critical importance of diversity, equity, and inclusion (DEI) for our society as a whole.

We will also examine the pivotal role of design as a professional discipline—one with its own unique set of tools, mindsets, and rich history—in advancing these principles. Toward the end of the chapter, we will explore the historical development of Inclusive Design, tracing its evolution from a niche concept to a foundational approach in modern practices. As we progress through the book, we'll see how DEI and design as a discipline can converge to create a transformative and powerful tool for shaping our future. Although awareness of DEI has grown significantly, the campaigns, actions, and pledges made in its name have also faced mounting criticism. Movements like Equitable Design offer a much-needed second breath to this effort, providing new ways to reinvigorate the push for a more inclusive and equitable world.

Understanding Inclusivity in Design

Settling into my chair, I couldn't help but reflect on my own privilege. I was a young, white, French man, highly educated and seemingly out of place in this discussion about inclusivity and equality. What right did I have to teach such a course? This thought gnawed at me, amplifying my anxiety. The question of legitimacy loomed large in my mind. Would the dean, then the students, and all the partners I planned on bringing on board find me credible for this role? Was I serving or undermining the cause? Was I occupying a space that should rightfully belong to someone whose voice needed to be amplified?

At the edge of any important pivot, imposter syndrome waits, growing louder as you approach the precipice of the unknown. This internal conflict was nothing new to me. Three years ago, I had embarked on a similar journey when I launched a program called Deplastify the Planet. Despite having a solid foundation in sustainability and systems thinking, I was stepping into uncharted territory. The responsibility of having my name associated with such a program was immense.

In the short term, it meant becoming the go-to person for any plastic-related inquiries on this side of the Bay Area. More significantly, in the long term, it meant influencing the minds of hundreds of students. These students, destined to become leaders and decision-makers, would carry the lessons they learned in my class into the world. The impact of this education would be profound and far-reaching.

Beside me was Ibrahim, one of my former students. He had been a key figure in my sustainable design class, and it was his passion and insight that had sparked the idea for this new course. Ibrahim was in his early 20s but had already made significant strides in the world of DEI with his startup. His presence was both a source of inspiration and a reminder of the impact one individual could make. His legitimacy and enthusiasm were the cornerstones of this project, and his support gave me the confidence to move forward.

But also, and ironically, his relative inexperience in teaching, navigating the professional world, and turning ideas into a viable business was, in fact, pivotal for me. Observing his journey, I saw a reflection of my own initial uncertainties and naivety concerning the topic of DEI. It became clear that entrepreneurship, whether it involves launching a business or developing new initiatives like this program, thrives on the convergence of individuals who possess both the intelligence to innovate and the audacity to take risks despite their lack of complete knowledge. It is this blend of ambition and ignorance, a willingness to venture into the unknown, that often paves the way for groundbreaking success.

As I sat there, the weight of the upcoming Zoom call settled in. This was the culmination of our initial efforts—dreaming, defining a vision, and our first steps in collaboration. We were on the brink of starting a new chapter, one that promised both challenges and opportunities. The agitation of the moment was palpable, yet it was also a reminder of the potential impact we could make.

Reflecting on our journey so far, I realized how much we had already accomplished. The countless hours of brainstorming, the discussions that stretched late into the night, the moments of doubt followed by sparks of inspiration—all had led us to this point. Each step had been a learning experience, shaping our approach and solidifying our commitment to the cause.

The significance of this moment wasn't lost on me. We could be transitioning from a phase of planning and vision-setting to one of execution and implementation. This transition was filled with uncertainty but also with the promise of real, tangible change. The journey ahead would undoubtedly be challenging, but it was a journey worth embarking on.

Solutions Lie with the Problem Creators

Just before the call, a crucial reminder surfaced in my mind. Despite my privileges, I was still the right person to advocate for change. The challenges of inclusion and equality are universal, requiring the active participation of everyone, victims of inequalities but also beneficiaries of privileges. For too long, society has viewed racism as a Black problem, placing the burden of resolution solely on the shoulders of Black individuals. Similarly, gender equality has been seen as a women's issue, with the expectation that women alone should fight for their rights.

However, it became clear to me that these issues cannot be effectively addressed by the marginalized alone. The burden of addressing these inequalities falls on those with privilege. It is our responsibility to recognize and analyze our advantages and then actively work to be part of the solution. This means challenging the status quo and using our positions of power and influence to create meaningful change.

There cannot be a viable and sustainable solution to exclusion and inequality if all parties are not involved. The most significant social justice victories in history have been achieved through the collective efforts of both the oppressed and their allies. It has been the silent majorities, those who may not face the injustices directly but are committed to ending unfair privileges, who have often tipped the scales toward progress.

As I sat there, the weight of this realization began to transform my anxiety into a sense of purpose. My role was not just to create a course but to use my platform to foster a deeper understanding of these issues among those who might not face them directly. It was about empowering everyone to see their part in the fight for equality and to encourage active participation in creating a more just world.

This realization reinforced the importance of the "Equitable Design" course. It wasn't merely an academic exercise but a call to action for students, faculty, and professionals to engage in the dialogue and work necessary to address systemic inequalities. The course would serve as a platform for diverse voices and perspectives, fostering an environment where real change could take root.

As the moments ticked by, I felt a growing sense of determination. I knew that my privilege did not disqualify me from this work; rather, it obligated me to it. I had the resources, the education, and the platform to make a difference, and it was my duty to use them wisely. This journey was about more than just a single course or a single pitch; it was about contributing to a broader movement for social justice and equality.

We All Have the Ability to Make a Difference

In the midst of the pandemic, the tragic events surrounding George Floyd's death sent ripples of realization through society. For many in the United States and around the world, it became clear that significant adjustments were needed in how we coexist and interact as a society. This period of awakening underscored the urgent need to improve our social interactions and strive for a more equitable coexistence.

My journey leading up to the concept of Inclusive Design began within the field of sustainability. A few years prior, I found myself grappling with the enormity of sustainability challenges. I realized that these issues required not only innovative solutions but also the active participation of everyone. My approach to contributing was to leverage the tools and resources I had at my disposal. As a citizen, I knew my choices—whether in consumption, communication, or educating my children—held the potential to make a difference. I aimed to be intentional in my efforts to address major challenges like climate change and, later, diversity and inclusion.

We often underestimate the power we wield as individuals within society. "Vote with your feet," the saying goes. In heavily commercialized societies, the way we allocate our spending and savings acts as a real support or rejection of political commitments. Frequently, I've spoken with executives of large organizations who express their fears about how their customers will react to their engaged initiatives.

The automotive industry is a prime example of this dynamic. Over the years, I have witnessed the entire value chain invest massively in creating a greener image. Although some might accuse these companies of greenwashing, the reality is that sustainability has climbed to the top of their agendas, backed by real investments and measurable impacts. This shift began with citizens demanding to reduce their personal footprints, which in turn pushed regulators in some countries and car manufacturers to offer greener solutions. In the meantime, the growing awareness of climate change's impact on our lives and the escalating costs of damage caused by global warming have created a profound sense of urgency and high stakes. What is particularly interesting about cars is that they serve as social representations of ourselves. In many countries, cars symbolize more than just a means of transportation; they are beacons of status, personality, and priorities. This demand cascaded across the value chain, prompting companies, often several degrees removed from the end users, to invest in innovative solutions to satisfy their clients.

An even more immediate manifestation of this power is evident in the plastic packaging and food industry. Eco-conscious consumers in large U.S. cities have been calling out retailers for the excess plastic packaging produced and used. By directing their purchases toward products with lighter or no packaging, these consumers have influenced the strategies of large multinationals, challenging the status quo. The collective actions of these individuals have pushed companies to reconsider their packaging methods and make substantial changes toward sustainability.

These examples illustrate the significant impact that individual choices can have on large-scale corporate strategies and lawmakers. By making conscious decisions about how we spend our money and what products we support, we drive industries to adopt more sustainable and responsible practices. It is a powerful reminder that each of us holds the potential to foster meaningful change within society.

In addition to being a citizen, I am a professional, a practitioner, and a teacher. Each of these roles allows me to make an impact through my work and actions, just as everyone has the capacity to do. I asked myself, given my skills, experiences, and network, how could I make a meaningful contribution? This question led to the inception of "Inclusive by Design."

At this point in my life, I had earned awards as a teacher in the #1 ranked public university in the world, UC Berkeley. I was the CEO of a trendy and fast-growing innovation hub in San Francisco, and I had hundreds of consulting projects under my belt, proving to top leaders that Design as a philosophy and a method could solve the most complex business challenges.

I believe that everyone possesses unique skills, abilities, and experiences in various fields. These capabilities, acquired over the course of our lives, can be channels for expressing meaningful intentions, such as making our society more inclusive. Whether you are an artist, an investor, a farmer, or a doctor, you always have the option to leverage your skills, your position, and your assets to contribute to a better society.

Artists, for instance, have historically used their work to take stands on important issues, creating engaged art pieces that inspire many supporters of important causes. From the Renaissance to modern times, artists have challenged societal norms and sparked crucial conversations through their creativity. Their work often serves as a catalyst for change, influencing public opinion and encouraging others to think differently about social justice issues.

Investors, too, can play a significant role in fostering inclusivity. By being intentional in how they allocate funds, they can support underserved communities and promote economic growth where it is most needed. Impact investing, for example, focuses on generating positive social and environmental outcomes alongside financial returns. This approach can help bridge economic disparities and provide opportunities for those who have been historically marginalized.

Farmers, often seen as the backbone of society's survival, have the power to effect change through their everyday practices. By choosing to hire lower-skilled employees, they can strengthen their communities and provide stable employment opportunities. Additionally, sustainable farming practices that care for the soil and environment ensure that future generations can continue to thrive. Farmers can advocate for fair treatment of workers and ethical production methods, contributing to social justice in the long term.

Doctors on the front lines of healthcare have a unique opportunity to make a profound impact. By providing culturally sensitive care, they can build trust with diverse communities and address healthcare disparities. Understanding and respecting cultural differences in healthcare practices can lead to better health outcomes and help reconcile communities with health institutions. This not only improves individual lives but also fosters a healthier, more equitable society.

These are just a few examples to illustrate that no matter what activities occupy the majority of our waking hours, we have the opportunity to contribute to a greater cause. Every profession, every skill set, and every position holds the potential for positive impact. By recognizing and harnessing this potential, we can make all those hours and efforts even more meaningful.

When we choose to use our talents and resources for the greater good, we become part of a larger movement toward inclusivity and equality. It's about seeing the bigger picture and understanding that our actions, no matter how small they may seem, can contribute to significant change. Each of us has the power to influence the world around us, and by doing so, we can help create a society that is more just, more inclusive, and more equitable for all.

The Power of Design and Intention

For over a decade, I have dedicated my career to making design more accessible and integral to strategic contexts for organizations, individuals, and students. My journey has been deeply intertwined with the evolving role of design in addressing complex challenges and driving meaningful change. This dedication has been rooted in a profound belief in the transformative power of design.

As an entrepreneur, I have come to realize the multifaceted power of design. From both a mindset and toolset standpoint, design can become a superpower when used properly. Designers possess a unique awareness of their environment and approach problems with humility and an almost naive perspective on new problems. This allows them to uncover new insights, embrace being proven wrong, and continually progress in their understanding of the systems they interact with.

One of the core strengths of designers is their willingness to be creative and try new things. They are not confined by conventional thinking and are always ready to think differently. This openness to innovation is coupled with a practical approach: designers regularly return to the field and engage with real-life situations to test and refine their ideas. This iterative process ensures that their solutions are grounded in reality and genuinely effective.

Moreover, what has always struck me about designers is their intentionality. They work with a clear purpose and vision. A team of designers at IBM once articulated this beautifully: "Design is the intention behind an outcome." This philosophy encapsulates the essence of design thinking. It is not just about aesthetics or functionality; it is about deliberate, thoughtful creation aimed at achieving specific, meaningful results.

I have been working for over a decade as both a strategic consultant and a college teacher, interacting every day with some of the most brilliant minds in the world. My task was primarily to demonstrate the power of a design philosophy and approach to solve complex problems; I did so by consulting with large organizations and being an ad hoc design lead on challenging projects. Trained as a mechanical engineer and a designer and very versed in business, I brought value to my clients by providing new insights into their challenges and offering creative yet feasible and profitable solutions.

At a pivotal moment in my career, I realized the potential of using my expertise in design to contribute to a more inclusive society. The question then became: how can design support inclusivity? How can the principles of design thinking—creativity, intentionality, humility, and practicality—be harnessed to foster a more equitable world?

This realization was not just a fleeting thought but a profound shift in my approach. It dawned on me that the same methodologies I had used to solve business problems and create sustainable solutions could be applied to the social realm. This was the genesis of my idea to develop a program that leverages design thinking to address issues of DEI.

The philosophy of Inclusive Design was nothing new and has been around with the idea of universal design since the mid-20th century. I refer to this remarkable history later in this chapter and feel very humble about following the track of remarkable thinkers and professionals. But times change, and there is still some room for designers like me to bring their own contributions and vantage points to the mix and make it a greater practice.

Of course, the path is not always clear and easy. It's natural to feel that your activities might be completely disconnected from DEI or to doubt your legitimacy and power to effect change. Many times, I have stumbled on compromises or found myself unable to seamlessly merge my tools with my pursuit of a more inclusive society. These obstacles can be discouraging and make the journey seem daunting.

However, this is where the power of intention becomes crucial. It's important not to aim for perfection or to get discouraged if you feel far from your goal. Staying true to your intention and motivation is essential. Share your goals and motivations widely because when you communicate your intentions, you invite collaboration and support. Solutions often arise from the collective effort and shared vision of a community, even if the path to those solutions is not immediately clear.

Maintaining this intentionality helps create a ripple effect. By consistently aligning your actions with your values, you inspire others to do the same. It's not about having all the answers or being in the perfect position to implement change from the start. It's about taking those first steps, no matter how small, and being open to learning and adapting along the way.

Remember, the impact of your efforts might not always be immediately visible. Walking the talk, embodying the principles of inclusivity and equity in your daily actions, already contributes to creating a more just society. Each step you take sets an example for others, demonstrating that change is possible and encouraging them to join you in your efforts.

In many ways, progress is about persistence and resilience. It's about facing setbacks and continuing to move forward, guided by your core values and the belief in a better future. Whether you're an artist, an investor, a farmer, or a doctor, your dedication to these principles can lead to meaningful changes in your field and beyond. This collective effort can truly go a long way in driving social change.

Ultimately, it's about realizing that every effort counts. By staying committed to your intentions and sharing your journey with others, you contribute to a larger movement. Together, these individual efforts accumulate, creating a powerful force for change. So, even if the path seems unclear or challenging, remember that your actions, rooted in genuine intention, are already making a difference.

I hope that everyone reading this book will reflect on these questions. Based on their roles as citizens, professionals, influencers, parents, or whatever positions they hold, how can they contribute to making our society more equitable? Each of us, through our unique capabilities and spheres of influence, can play a part in driving change. This collective effort can truly go a long way.

The journey toward inclusivity is not a solitary one. It requires the collective effort of individuals from all walks of life, leveraging their unique skills and perspectives to create a more just and equitable world. As you read this book, I encourage you to think about your own journey. Consider how you can use your abilities to contribute to this cause. Remember, the power of design lies not just in its ability to create beautiful and functional solutions but in its potential to drive meaningful, lasting change in our society.

Finally, the Zoom call began, and the smiling face of the senior vice-dean appeared on my screen. Her presence radiated both warmth and authority, instantly putting me at ease. As we delved into our conversation, the initial tension melted away, replaced by an enthusiastic and passionate discussion. We explored the potential impact of the proposed class, shared visions of a more inclusive educational environment, and brainstormed ways to bring the concept to life.

The senior vice-dean's supportive nods and insightful contributions fueled my excitement and confidence. Our dialogue felt less like a pitch and more like a collaborative effort to build something meaningful. By the end of the call, we had not only agreed to launch the new class but also laid the foundation for what I realized would become a flagship initiative of my career: Inclusive by Design, Crafting Products and Services for a More Equitable World.

This project, born out of a simple idea and a crucial conversation, has now blossomed into a lifelong commitment. The journey from my initial hesitation to this moment of triumph encapsulates the essence of what Inclusive by Design stands for—a continuous pursuit of creating a more equitable world through thoughtful and inclusive design. As I logged off the Zoom call, I felt a renewed sense of purpose, ready to embark on this transformative journey.

The Importance of Diversity, Equity, and Inclusion

In the wake of George Floyd's tragic death, a spotlight was cast on the critical importance of DEI. It was remarkable to witness a surge of commitment from leaders across both the public and private sectors, all recognizing DEI as a crucial priority. Political figures and corporate executives alike vowed to invest heavily in creating meaningful changes within their organizations and communities. The pledges were grand: millions of dollars allocated to diversity initiatives, the establishment of numerous support groups, and a renewed focus on fostering inclusive environments.

First and foremost, it's important to acknowledge the positive momentum that emerged. The widespread reaction from society, especially those who chose to speak out and act, was overwhelmingly positive. This collective awakening to the importance of DEI represented a significant step forward. It underscored a genuine desire among many to address systemic inequalities and to foster more inclusive environments.

However, on closer examination, a disconnect becomes evident between the funds pledged and the funds actually spent. Many companies, despite their well-publicized commitments, fell short in following through with their financial promises. The initial enthusiasm for DEI often did not translate into substantial, long-term investments.

And years later, a lot of influential voices are finally asking what exactly this DEI industry is and where the significant, deep, long-term effects are of this abundance of declarations. The promises haven't been achieved, and somehow, the DEI wave has backfired, provoking frustration and deception among the best supporters of the cause, those who aspire to a more equitable workplace but have grown suspicious of the DEI bureaucracy, weary of the training seminars, inured to the formulaic language, and uncertain about how to navigate what may appear to be the competing claims of "talent" and "diversity."

Corporations pledged billions of dollars toward DEI initiatives following the murder of George Floyd in 2020, but the actual impact of those commitments has been limited so far. Here are some key statistics and findings:

- Companies pledged a combined $340 billion toward racial equity after Floyd's death, with many commitments coming from the financial sector.[1] However, only $250 million has been spent toward that specific initiative as of 2023, according to Creative Investment Research.[2]

- A survey by CNBC and the Executive Leadership Council found that 88% of Black executives reported that their companies made DEI commitments in 2020.[3] However, nearly one-quarter (23%) felt they were not equitably compensated relative to their peers, and about two-thirds reported under-representation of Black employees in upper management.

- Although 41% of companies increased Black executive representation on their senior leadership teams after 2020 and 40% tied DEI goals to compensation for senior leaders, there is still a significant opportunity gap for Black professionals.

- Job listings for DEI roles surged by 123% in the three months after Floyd's death as companies rushed to hire DEI professionals.[4] However, DEI roles were down 19% in 2022 compared to 2020, with major tech companies like Twitter, Meta, and Amazon cutting DEI staff.

- The attrition rate for DEI roles was 33% at the end of 2022, compared to 21% for non-DEI roles, indicating a disproportionate impact on DEI professionals during layoffs.[5]

Although corporations made public commitments and financial pledges toward racial equity and DEI after the events of 2020, the data suggests that the actual implementation and impact of these initiatives have been limited so far. Many Black professionals still face underrepresentation, pay disparities, and a lack of sustained commitment from their employers.

Moreover, the allocation of these resources frequently missed the mark. In many cases, companies hired significant numbers of people of color but often placed them in lower-paying positions with limited responsibilities. This superficial approach to diversity failed to address the deeper issues of equity and inclusion within the organizational structure. Although the hiring statistics looked impressive on the surface, they did not reflect true integration or empowerment of these employees.

Another overlooked aspect is the high churn rate in these roles. Many organizations did not publicly disclose the turnover rates among their newly hired diverse workforce. The lack of retention efforts and support systems meant that many individuals left their positions, disillusioned by the lack of genuine opportunities for growth and advancement.

With my experience interacting with representatives of Fortune 500 companies and business leaders, I feel like the desire to create a meaningful impact was genuine. Yet the structure of businesses made it easier for them to follow the paths of least resistance: finding and harvesting the low-hanging fruit, and tasking ad hoc teams to make quick and affordable actions to gather some numbers to be shown inside and outside the organization. I like to think that courage and determination were not lacking in the board rooms but that the effort was dictated by short-term results, peer pressure, and preservation of the business, which suggest limited risks and visible and immediate results that got in the way of bold, systemic changes. Overall, I join those who think this unprecedented DEI campaign was disappointing in its actual results and follow-up. Yet I think it created a form of cultural revolution that makes it easier for more long-term and fundamental actions like Inclusive Design to exist and spark in the future.

In addition to these explanations, I also believe that, in certain instances, leaders and decision-makers were simply not aligned with or supportive of some of these changes. In the United States, a significant number of individuals either harbor fear about the rise of specific communities or benefit from maintaining the current power structures, leading them to make conservative and protective decisions. The country is undeniably polarized, and the concept of equality varies widely depending on the perspective and background of the person you're speaking with. In this context, it's crucial to acknowledge that creating a more inclusive and equitable society will require bridging divides, engaging with people across a spectrum of beliefs, and navigating numerous challenges along the way.

This highlights a critical flaw in the approach to DEI: it cannot be achieved through superficial measures or token hires. True diversity and inclusion require comprehensive strategies that address systemic issues, provide equitable opportunities, and foster an environment where all employees and customers feel valued and supported. It's not just about bringing diverse individuals into the organization but also about ensuring they have a voice, influence, and a clear path for development.

It's also crucial to recognize that DEI should not be reduced to mere numbers or compliance projects. It must be deeply ingrained in the organizational culture, extending far beyond what is easy to measure, such as workforce diversity. Real progress involves a holistic approach, integrating DEI principles into every aspect of an organization's operations and strategy.

The Immediate Benefits of DEI for Organizations

There have been numerous instances where genuine DEI efforts have led to positive outcomes for both businesses and society. For example, diverse teams have been shown to drive innovation and better decision-making, ultimately leading to improved financial performance. A McKinsey study found that companies in the top quartile for ethnic and cultural diversity on executive teams were 36% more likely to outperform their peers in terms of profitability. This illustrates the tangible benefits of a truly inclusive approach.[6]

Yet the challenge remains to go beyond initial efforts and become more intentional and systematic in our approach to inclusion and equality. The events following George Floyd's death have provided important lessons. We must move toward systemic change, ensuring that DEI initiatives are not just reactive measures but proactive strategies embedded in the fabric of our organizations.

But the impact of inclusive practices that aim for more equitable outcomes is far greater than this.

And although in the United States, DEI tends to be associated with racial justice or LGBTQ+ rights, there is much more to think about. Inclusive and Equitable Design aims at increasing the quality of experience for people with physical disabilities; contributing to gender inclusion; improving products, services, and experiences for people with neurodivergence; offering better support and services to small businesses; easing the access to funding to underrepresented entrepreneurs; bridging the digital gap for older adults; and much more. The scope is endless, leaving space for anyone to contribute and find their area of interest and personal connection with the challenge.

Let me tell you how a more diverse team can save a major business in challenging times.

As I settled into the call with Nadine, the head of DEI for a large real estate group, I could sense the weight of her responsibilities through the screen. We had been chatting for about 15 minutes, discussing the typical challenges faced in implementing DEI initiatives in the real estate industry. The conversation flowed naturally, touching on the common hurdles and the relentless pushback that often accompanies efforts to foster inclusivity in traditionally homogeneous sectors.

"Real estate is such a tough industry for DEI," Nadine remarked, her voice carrying the fatigue of countless similar conversations. "It's been an uphill battle to convince the leadership that diversity isn't just a checkbox but a strategic necessity."

I nodded, understanding the frustration. Then Nadine's tone shifted as she began to share a story that had evidently had a profound impact on her and the company.

"Let me tell you about something that happened during the pandemic," she started, her eyes lighting up with a mix of pride and astonishment. "Our company was struggling to navigate the sudden shifts in market demand. We were all caught off guard, trying to figure out how to adapt to the new normal."

She paused, gathering her thoughts before continuing. "We have a diverse team of brokers, which was a result of a deliberate effort to embrace different perspectives and backgrounds. I knew that these varied experiences and connections would be valuable, but it wasn't until the pandemic hit that we fully realized the extent of their impact. These brokers brought perspectives and connections that we hadn't fully appreciated until everything changed overnight."

Nadine explained how the pandemic had drastically altered the business landscape. Companies were scrambling to meet the surging demand for local warehouses and logistics platforms as e-commerce skyrocketed. Traditional market segments and prime real estate locations suddenly took a backseat to more strategic yet previously overlooked areas.

"Our brokers who came from more diverse backgrounds had deep ties to suburban and underserved urban communities," Nadine explained. "They understood these markets in ways our traditional strategies had never considered. When CBD office spaces became ghost towns, these brokers identified opportunities in underappreciated yet strategically positioned properties. They turned these locations into warehouses and distribution centers, perfectly suited for the burgeoning delivery business."

She leaned forward, emphasizing the turning point. "These insights were game-changing. Properties in areas we had never prioritized suddenly became our lifeline. Their knowledge and connections allowed us to pivot quickly, securing leases and purchases that kept us afloat and competitive during the toughest months of the pandemic."

The realization among the company's executives was striking. What had once been seen as secondary markets or overlooked neighborhoods now represented critical assets in their portfolio. The diverse workforce, with their unique perspectives and networks, had not only helped the company survive but thrive in a crisis.

"This wasn't just about having a more creative or problem-solving team," Nadine continued. "It directly translated into immediate revenue opportunities and long-term resilience. Investing in diversity, equity, and inclusion proved to be not just a moral or ethical decision but a strategic business advantage."

Nadine's story encapsulated the essence of why DEI initiatives are so crucial. It wasn't just about ticking boxes or improving company culture—it was about unlocking new opportunities and ensuring the company's ability to adapt and succeed in a rapidly changing world. Diversity, she highlighted, was a source of strength and resilience, providing tangible business benefits that went beyond the usual narratives.

As we wrapped up our call, I couldn't help but feel inspired by Nadine's experience. It was a powerful reminder that embracing diversity isn't merely an aspirational goal; it's a strategic imperative that can transform challenges into opportunities and pave the way for sustainable success.

This story is no exception. Throughout this book, we will see countless illustrations of how an intentional and systemic effort toward DEI contributes to better business and increases happiness in our communities.

Beyond employee diversity, there lies an immense and exciting source of opportunities on the customer-facing side of an organization. Embracing inclusiveness in products, services, and processes can unlock a wealth of possibilities and benefits, propelling any company to new heights.

Investing in Equitable Design and prioritizing DEI when developing products and services can be highly rewarding from a business standpoint. These efforts not only foster a more inclusive and supportive work environment but also translate into tangible business benefits.

We can highlight three main buckets of financial upside for businesses engaging in Equitable Design:

1. Broadening the initial audience, increasing the serviceable market

2. Mitigating risks and improving resilience, limiting exposure to loss of business and reputation

3. Creating a competitive advantage in their market and overperforming the competition

Broadening the Audience

Broadening the audience through Equitable Design is a compelling reason for businesses to invest in DEI. When products and services are designed to be more inclusive, they naturally appeal to a wider range of users, thus expanding the potential customer base. This approach can be understood in several key ways.

First, reducing barriers to access and use is crucial. Many products and services are not accessible to people with disabilities, such as those with visual impairments. By designing with accessibility in mind, businesses can capture an untapped market. For instance, incorporating features like screen reader compatibility or tactile feedback can make products more usable for visually impaired individuals, thereby increasing the user base.

In a lot of cases, those audiences might be larger than one would think. Based on the most recent data, over 7 million Americans aged 40 and older have vision impairment, with 1 million of them being classified as legally blind, and those numbers are expected to double by 2050.[7, 8]

Another important aspect to consider is the changing ethnic demographics of the United States. According to studies, the country is projected to become a majority-minority nation by 2045. This means that more than 50% of U.S. citizens will be from current racial or ethnic minority groups. This demographic shift highlights the growing importance of designing products and services that cater to what are currently considered racial minorities. A report by the U.S. Census Bureau projects that in 2060, non-Hispanic whites will account for less than 44% of the US population, whereas Hispanic, African American, Asian, and other racial and ethnic groups will make up the majority.[9]

Geographical accessibility is another important factor. Ensuring that products and services are available in more locations allows more people to access them. This is particularly relevant for businesses with physical stores or services that require proximity. Expanding the reach by positioning stores in more diverse and underserved areas can significantly increase the number of potential customers.

Beyond physical and functional accessibility, fostering a sense of belonging is crucial. This involves creating culturally relevant marketing campaigns and products that resonate with different communities. For example, branding and product features that reflect the values and aesthetics of diverse cultural groups can make these products more appealing. In the United States, which is on the brink of becoming a majority-minority country, it is increasingly strategic for companies to engage with and serve these diverse communities effectively.

By focusing on inclusivity, businesses not only fulfill ethical and social responsibilities but also unlock significant economic opportunities. Embracing diversity allows companies to better serve a broader audience, thus driving growth and ensuring long-term sustainability in an increasingly diverse market.

Investing in DEI is not just a moral or ethical imperative; it also plays a crucial role in mitigating risks and improving business resilience. Here are the key aspects to consider.

Mitigating Reputational Risks

In today's interconnected world, where information spreads rapidly via social media, companies that fail to actively engage in DEI efforts risk facing significant reputational damage. Although most companies may not be deliberately exclusionary, the absence of proactive DEI initiatives can be perceived negatively by both internal and external stakeholders. Employees and consumers are increasingly valuing corporate social responsibility, and companies that neglect DEI may be viewed as out of touch or indifferent to important social issues.

Instances of backlash against companies perceived as insensitive or noninclusive have shown how damaging negative publicity can be. These reputational risks can lead to loss of consumer trust, boycotts, and a decline in employee morale and retention. On the other hand, companies that are seen as leaders in DEI can build stronger, more positive relationships with their customers and workforce, enhancing their overall brand reputation.

The Dolce & Gabbana scandal in China began in 2018 when the brand released a series of promotional videos that were widely criticized as culturally insensitive. The ads depicted a Chinese model struggling to eat Italian food with chopsticks, accompanied by a patronizing voiceover. The backlash intensified when alleged derogatory messages from cofounder Stefano Gabbana surfaced, further alienating Chinese consumers. The fallout was severe: the brand's Shanghai fashion show was canceled, Chinese celebrities severed ties, and D&G products were removed from major Chinese e-commerce platforms.

The long-term impact has been profound. Dolce & Gabbana's reputation in China, a critical market for luxury brands, remains deeply tarnished. The brand has seen a reduction in its physical presence in China, with several boutiques closing. Additionally, it has been frozen out of major online retail platforms like Tmall and JD.com, severely limiting its ability to engage with Chinese consumers. This scandal serves as a stark reminder of the immense financial and reputational risks associated with failing to adopt Equitable Design practices that respect and understand cultural sensitivities.[10]

The incident underscores the necessity for companies to engage in Equitable Design practices that take into account the cultural contexts of the markets they serve. Without this awareness, brands risk alienating key consumer bases, leading to significant financial losses and long-term damage to their global reputation.

Enhancing Business Resilience

A diverse and inclusive workforce brings a variety of perspectives and experiences, which can be invaluable in identifying risks and developing innovative solutions. This diversity of thought is crucial for problem-solving and adapting to changing market conditions. By fostering an inclusive environment, companies can tap into a broader range of ideas and insights, making them more adaptable and resilient.

Furthermore, having a diverse client base and supplier network reduces the risk of over-reliance on a single market segment or source. This diversification is critical in protecting the business from geopolitical, economic, or social disruptions. For example, if a company relies heavily on suppliers from a single region and that region faces political instability, the company's operations could be severely impacted. However, a diversified supplier base mitigates this risk, ensuring greater stability and continuity.

Although the primary motivation for DEI should be a genuine commitment to creating a fair and inclusive environment, the business benefits are significant. By reducing reputational risks and enhancing resilience through diversity, companies position themselves for long-term success and sustainability. This strategic approach to DEI is not just beneficial but essential in today's rapidly evolving business landscape.

Creating a Competitive Advantage in the Market

Intentional investment in DEI not only mitigates risks but also creates strategic advantages. Companies that prioritize DEI can leverage their diverse workforce to better understand and serve a wider array of customer needs. This can lead to increased market share and loyalty, particularly in a global marketplace that is becoming more diverse.

Investing in DEI and Equitable Design can also provide a significant competitive advantage in the market. By tailoring products and services to meet the needs of often-overlooked audiences, companies can tap into new customer bases and differentiate themselves from competitors. This approach is particularly effective in commoditized markets, where differentiation can be challenging.

Many businesses focus on the largest, most apparent segments of the market, often overlooking specific groups with unique needs. By identifying and addressing these needs, companies can create products that resonate deeply with these underserved audiences. For example, designing products that cater to individuals with neurodivergence, who make up about 20% of the U.S. population, can open up substantial new market opportunities. Features that improve accessibility or user experience for neurodivergent individuals can attract this significant demographic, setting a company apart from its competitors, which may not have considered these specific needs.

This strategy not only broadens the customer base but also fosters brand loyalty. When consumers see that a company genuinely cares about their needs and goes the extra mile to accommodate them, they are more likely to remain loyal to that brand. This sense of belonging and appreciation can translate into long-term customer relationships and positive word-of-mouth marketing.

Additionally, inclusive products often have broader appeal beyond their initial target audience. Features designed to assist neurodivergent individuals, for example, can also benefit a wider range of users who appreciate thoughtful, user-friendly design. This can enhance the overall user experience and attract a more diverse customer base.

In highly competitive and commoditized markets, such differentiation is invaluable. It provides a clear reason for consumers to choose one brand over another, helping companies stand out in a crowded marketplace. By prioritizing DEI and Equitable Design, businesses not only fulfill a social responsibility but also gain a strategic advantage that drives growth and success.

Investing in DEI and Inclusive Design practices is not just an ethical choice; it is a strategic business decision that can lead to expanded market reach, increased customer loyalty, and a strong competitive position.

The Long-Term Risks of Growing Inequality: A Path to Societal Collapse

In recent years, the discourse around DEI has highlighted the immediate business and social benefits of investing in equitable practices. However, it is crucial to recognize that beyond these immediate impacts, there are significant long-term stakes involved. Just as climate change poses a major threat to the future of humanity, an inequitable society holds the potential for similarly catastrophic consequences.

It has become increasingly urgent to prioritize the fight for social justice, inclusiveness, and equality within our society and communities. Failing to do so risks plunging our civilization into profound imbalance, potentially leading to violence and societal collapse. Numerous examples of such dangers can be found in the past of our civilization, and current trends offer alarming evidence that we are heading toward a major catastrophe, much like the one posed by climate change. Addressing these issues now is essential to prevent further destabilization and ensure a sustainable future for all.

Historical Lessons on Inequality and Collapse

History provides numerous examples where unchecked inequality led to societal collapse. The fall of the Roman Empire, driven by economic disparity and social unrest, is a classic example. The French Revolution, sparked by the extreme wealth divide between the nobility and the common people, is another. More recently, the Arab Spring demonstrated how economic disenfranchisement can ignite widespread upheaval.

The Fall of the Roman Empire

The Roman Empire, once a powerful civilization, began to decline in the 3rd century AD due to severe economic disparity and social unrest. Wealth became concentrated among a small elite, while the majority of the population lived in poverty. This imbalance weakened the empire's cohesion, making it vulnerable to external pressures. By 476 AD, the Western Roman Empire fell, marking the end of a centuries-long era of Roman dominance. Economic disparity and the neglect of the common populace played significant roles in this downfall.

The French Revolution (1789–1799)

The French Revolution was ignited by extreme inequality between the wealthy aristocracy and the struggling common people. By the late 18th century, France's finances were in shambles, and heavy taxation fell disproportionately on the Third Estate, comprising commoners. Meanwhile, the nobility and clergy enjoyed vast wealth and privileges. This stark divide led to widespread anger and the eventual uprising against King Louis XVI and the aristocracy. The revolution led to the abolition of the monarchy, the rise of Napoleon Bonaparte, and profound changes in the social and political fabric of France. The revolution's slogan, "Liberty, Equality, Fraternity," symbolized the people's demand for justice and equal rights.

The Arab Spring (2010–2012)

More recently, the Arab Spring, a series of anti-government protests across the Middle East and North Africa, highlighted the dangers of economic disenfranchisement and political repression. Beginning in Tunisia in late 2010, widespread discontent over unemployment, rising prices, and political corruption led to protests that toppled long-standing regimes, including that of Egyptian President Hosni Mubarak in 2011. The movement spread to countries like Libya, Syria, and Yemen, resulting in varying degrees of political upheaval. Economic inequality, particularly the lack of opportunities for young people, was a driving force behind the protests. The Arab Spring underscored how economic and social grievances can ignite widespread unrest, leading to significant and sometimes violent change.

These historical events illustrate the crucial importance of addressing inequality. When societies fail to maintain balance and justice, they risk falling into chaos and collapse.

Jason Hickel, in his work *Less Is More* (Windmill Books, 2021), discusses how the transition from feudalism to capitalism during the early modern period in Europe laid the groundwork for systemic inequality. The enclosure of commons and the commodification of resources prioritized the wealth accumulation of a few over the well-being of the many. This historical pattern illustrates the long-term destabilizing effects of prioritizing economic growth and wealth for the elite over equitable resource distribution.

The Importance of Addressing Inequality to Prevent Societal Collapse in the Context of Climate Change

Recent research underscores the intertwined risks of climate change and economic inequality as dual threats that could precipitate societal collapse. Historical and contemporary analyses reveal how environmental stresses and social vulnerabilities interact to destabilize societies.

The historical analysis conducted by Daniel Hoyer and colleagues underscores the critical role that social cohesion and equitable resource distribution play in a society's resilience to environmental shocks. The study's findings show that societies with these qualities have consistently fared better during periods of environmental stress. On the other hand, those marked by deep inequalities and internal conflicts were much more susceptible to collapse.[11]

The examples of the Qing Dynasty and the Ottoman Empire show how much granary systems, resource distribution systems, intricate irrigation systems, and robust social welfare allowed those societies to sustain major drought and famine. But when corruption and elite competition arose, the state's capacity to manage crises deteriorated, ultimately toppling the dynasty.

These historical examples are not just relics of the past; they provide critical lessons for contemporary society. As we face the dual challenges of climate change and increasing social inequality, the importance of fostering social cohesion and equitable resource distribution cannot be overstated. The interaction between environmental stressors and social structures is complex, but the message is clear: societies that prioritize inclusivity, equity, and cooperation are more likely to withstand and adapt to the challenges posed by a changing climate.

Inequality affects more than just economic metrics; it erodes social cohesion and trust. High levels of inequality correlate with increased crime rates, poorer health outcomes, and lower educational attainment. These factors contribute to a less stable and more fragmented society. Hickel argues that the social contract is weakened in highly unequal societies, leading to decreased civic engagement and heightened political polarization.

Environmental degradation driven by unchecked economic growth disproportionately impacts the poor, further entrenching inequality. This creates a feedback loop where environmental harm and economic disparity reinforce each other, posing a dual threat to societal stability.

In the context of modern society, this underscores the importance of DEI as a foundational pillar for resilience. Just as climate change represents a formidable challenge, so too does the growing disparity within and between societies. Without addressing these disparities, the compounded effects of inequality and climate change could lead to social unrest, economic instability, and even societal collapse.

The lessons from history make it evident that Equitable Design and inclusive practices are not just ethical imperatives but strategic necessities for ensuring the long-term survival and thriving of societies. As we move forward, integrating DEI principles into all aspects of societal development—be it in governance, business, or community planning—will be crucial in building the resilience needed to navigate the complex challenges of the 21st century.

Why Design?

In light of those events, combined with the now well-known effects of climate change, we should feel a sense of urgency to fight against inequalities and contribute to tighter and more distributive societies. The usual reflex for this would be to turn to politics and make this an institutional challenge, ultimately putting the burden on the state. My bet is to show that this is not the only way and that design as a discipline, as a business imperative, can be a complementary direction to address this pressing issue.

In this book, when I refer to Design, I am speaking about the discipline in its broader sense—the comprehensive process of conceiving, creating, and refining a product or service to meet user needs, solve problems, and enhance experiences. This goes far beyond the visual rendering or aesthetic appeal, which is often the first association people make when thinking of design. Although the visual component is a crucial aspect of design, the discipline itself encompasses understanding users, researching behaviors, prototyping, testing, and iterating to create solutions that are both functional and meaningful.

This distinction is vital because Design as a discipline is about shaping the way people interact with the world and addressing systemic issues, not just making things "look good." For instance, a beautifully designed app interface is of little value if it's inaccessible to certain users due to poor functionality or exclusionary features.

The Foundational Role of Design in Tackling DEI Challenges

Design is a fundamental starting point for creating any product, service, or system. Everything we interact with daily, from the devices we use to the spaces we inhabit, has been designed with a specific intention and process. This underlying principle highlights the immense power of design in shaping our world and, consequently, the societal impacts it can have.

Design as the Beginning of All Creation

Every product and service starts with an idea originating in someone's mind. This idea is driven by the need to solve a problem, fulfill a desire, or improve an experience. This initial spark is where design begins. Whether consciously acknowledged or not, the process of conceptualizing, planning, and crafting a solution is inherently a design activity. Even when done without formal training or awareness, this phase involves critical design thinking: identifying needs, brainstorming ideas, and outlining how to bring those ideas to life.

Design, at its essence, is about creating intentional and functional solutions. The importance of this initial phase cannot be overstated because it sets the direction and intention for the entire development process. If we aim to address societal challenges, particularly those related to DEI, embedding these principles at the very beginning of the design process is crucial. By starting with inclusive intentions, we can ensure that the final products and services are not only functional but also equitable and accessible.

The Pervasiveness of Design

The impact of design extends far beyond the initial idea. Look around, and you will see that every object, interface, and environment has been (carefully) designed. From the devices we use daily to the public spaces we navigate, each has undergone a design process considering usability, aesthetics, functionality, and user experience. This ubiquity of design underscores its importance in shaping our interactions, behaviors, and overall quality of life.

Design influences everything we touch and see, often without us realizing it. Effective design can simplify tasks, enhance user satisfaction, and improve accessibility. Conversely, poor design can lead to frustration, exclusion, and inefficiency. Recognizing the pervasive nature of design highlights its potential as a powerful tool for fostering inclusivity. By making conscious design choices that prioritize DEI, we can create environments and products that serve a broader spectrum of people, addressing their diverse needs and preferences.

The Storytelling Power of Design

Another crucial aspect of design is its ability to tell stories. Good design is not just about functionality and aesthetics; it's also about conveying a narrative. Henry Ford once said, "Every object tells a story." This storytelling element can imbue objects with deeper meaning and connection to their users.

Consider the example from the documentary *Objectified* (by Gary Hustwit, 2009), where a designer explains the thoughtful design of Japanese toothpicks. The ends of these toothpicks are designed to be broken off, serving two purposes: it indicates that the toothpick has been used, and it acts as a small rest to keep the other end sanitary. This story illustrates how a simple object can carry significant cultural meaning and practical thoughtfulness, which is well-known in Japan but less so in other parts of the world. This example demonstrates how design can embed cultural values and practices into everyday objects, making them more meaningful and functional.

Another illustrative example is the design of the original iPhone. When Apple introduced the iPhone in 2007, it wasn't just a piece of technology; it was a revolution in how we interact with our devices. The design told a story of simplicity and elegance, encapsulated in a single home button and a multitouch screen that invited users to explore through touch. This design narrative was about breaking away from the cluttered and complex interfaces of the past and moving toward a more intuitive and user-friendly experience. The story of the iPhone's design helped to create a new standard in the tech industry, emphasizing user-centric design and aesthetic simplicity.

The storytelling power of design is particularly relevant in the context of DEI. It enables communication with a diverse audience, creating a sense of belonging and addressing cultural nuances. By telling stories through design, we can connect with people on a deeper level, fostering empathy and understanding. This capability is vital for addressing DEI challenges, as it helps bridge cultural gaps and create products and services that resonate with a wider audience.

The visual design of Google's Doodles often reflects diverse cultural celebrations and historical events, telling stories that resonate with different communities around the world. By incorporating these elements into its design, Google connects with users on a cultural level, fostering a sense of inclusion and representation.

Many examples highlight how design can be a powerful tool for storytelling, conveying messages that go beyond the surface level. In the context of DEI, this storytelling aspect of design becomes a bridge that connects different cultures, perspectives, and experiences. It allows designers to create products that are not only functional and aesthetically pleasing but also deeply meaningful and resonant with a diverse audience.

By leveraging the storytelling power of design, companies can create products that speak to the values and needs of various communities. This approach fosters a sense of belonging and inclusion, helping to address DEI challenges and create a more equitable society. Through thoughtful and intentional design, we can tell stories that inspire, connect, and empower people from all walks of life.

Systems Thinking

Designers often employ systems thinking, a methodology that involves understanding the interconnected components of a problem and how they influence one another within a whole. This approach is crucial for addressing DEI challenges because it allows for a comprehensive examination of the root causes of inequality rather than merely addressing symptoms.

Systems thinking in design helps identify the various elements that contribute to a problem and their interdependencies. For example, when tackling accessibility in urban spaces, systems thinking considers not just the physical design of buildings and public spaces but also the policies, social attitudes, and economic factors that influence accessibility. This holistic view enables designers to create more effective and sustainable solutions that address the underlying issues of inequality.

By viewing problems through a systems lens, designers can anticipate the potential consequences of their solutions and make adjustments accordingly. This proactive approach is essential for DEI, as it ensures that interventions do not inadvertently create new barriers or exacerbate existing inequalities. Systems thinking encourages a deep understanding of the entire ecosystem in which a design solution will operate, leading to more inclusive and equitable outcomes.

Additionally, systems thinking helps avoid the creation of new or bigger problems than the ones being solved. Narrow and exclusive approaches can often lead to unintended consequences, where the resources and outcomes of a solution inadvertently cause new issues. An example of this is the rebound effect, a well-known phenomenon in environmental economics.

The rebound effect occurs when the gains from improved efficiency in resource use are offset by increased overall consumption. For instance, if a new, more energy-efficient heating system is installed in homes, people may use the heating system more liberally because it costs less to run, ultimately leading to no net reduction in energy consumption or even an increase. This is a classic case where a narrowly focused solution aimed at reducing energy use leads to a counterproductive outcome due to not considering the broader system.

In the context of DEI, a similar situation could arise if a company implements a policy to increase diversity by hiring a significant number of people from underrepresented groups but fails to address the workplace culture and support systems. Without an inclusive environment, these new hires may face challenges that lead to high turnover rates, perpetuating a cycle of recruitment without retention. Thus, the initial problem of lack of diversity remains unsolved, and additional resources are wasted.

By adopting systems thinking, designers and organizations can better understand the bigger picture and the potential ripple effects of their actions. This comprehensive approach allows for the identification and mitigation of negative externalities, ensuring that the solutions implemented do not create additional problems. Systems thinking therefore fosters the development of sustainable, inclusive, and effective DEI strategies that consider all aspects of the ecosystem in which they operate.

Multidisciplinary Integration

Design integrates insights from various disciplines, including psychology, sociology, technology, and more. This multidisciplinary approach is particularly beneficial in addressing the complex, multifaceted nature of DEI challenges. Each discipline brings a unique perspective and set of tools that can enhance the design process.

For instance, incorporating psychological principles helps designers understand user behavior and motivations, leading to more intuitive and user-friendly products. Sociological insights can illuminate how social structures and cultural norms impact user experiences, ensuring that designs are culturally sensitive and inclusive. Technological advancements provide innovative tools and platforms that can make products and services more accessible.

The integration of these diverse perspectives fosters a richer and more comprehensive design process. It allows designers to tackle DEI challenges from multiple angles, ensuring that solutions are robust and well-rounded. This multidisciplinary approach also promotes collaboration and knowledge sharing, further enhancing the effectiveness of design interventions.

In the context of DEI, multidisciplinary integration ensures that solutions are not only innovative but also grounded in a deep understanding of the diverse factors that influence inequality. By drawing on a wide range of expertise, designers can create more holistic and impactful solutions that address the complex realities of the communities they serve.

However, multidisciplinary integration is not limited to academic disciplines. It also encompasses a wide range of life experiences, education outside of formal schooling, diverse walks of life, ages, and more. This broader definition of diversity can bring even richer insights and more innovative solutions to the table.

For example, consider the inclusion of individuals from different age groups in the design process. Older adults may provide valuable insights into the usability needs of senior citizens, and younger individuals can offer fresh perspectives on emerging trends and technologies. This intergenerational collaboration can lead to products and services that are accessible and appealing to a wider demographic.

Similarly, incorporating the experiences of people from various socioeconomic backgrounds can reveal unique challenges and opportunities that might otherwise be overlooked. For instance, a person who has lived in a low-income neighborhood can provide firsthand knowledge of the barriers faced by underserved communities, guiding the design of more effective and equitable solutions.

Moreover, individuals with nontraditional educational backgrounds or diverse career paths can contribute innovative ideas and approaches. A self-taught programmer, an artist, or a community organizer may bring unconventional yet highly valuable perspectives to the design process, challenging traditional methods and inspiring new ways of thinking.

By embracing this broader definition of multidisciplinary integration, designers can ensure that their solutions are not only academically informed but also deeply rooted in the real-world experiences of diverse populations. This holistic approach to design fosters creativity, inclusivity, and resilience, ultimately leading to more effective and sustainable DEI initiatives.

Incorporating these diverse experiences and perspectives enriches the design process, making it more inclusive and comprehensive. It ensures that the solutions developed are not only theoretically sound but also practically applicable and culturally resonant, effectively addressing the multifaceted nature of DEI challenges.

Creating Meaningful Change Through Design

If we aim to create meaningful societal change, starting with design is a strategic and effective approach. By embedding principles of DEI into the design process from the outset, we can ensure that the resulting products and services are accessible, inclusive, and equitable. This proactive integration of DEI can help address systemic inequalities and create solutions that serve a broader and more diverse audience.

For instance, when designers prioritize accessibility, they create products that are usable by individuals with disabilities, thereby expanding the user base and promoting inclusivity. When cultural sensitivities and diverse perspectives are considered, products and services resonate more deeply with varied communities, fostering a sense of belonging and representation.

The Ripple Effect of Intentional Design

Starting with design allows us to influence the entire lifecycle of a product or service. By setting inclusive intentions at the beginning, we can shape development, marketing, distribution, and user-engagement practices. This holistic impact underscores the strategic importance of design in achieving DEI goals.

As an example, consider the design of urban spaces. Inclusive Design principles can ensure that public areas are accessible to everyone, regardless of physical ability, age, or socioeconomic status. This approach not only improves the quality of life for all residents but also promotes social cohesion and equity.

Design as a Personal and Professional Tool

Design has always been more than just a profession for me; it has been the most powerful tool I have had available throughout my career. Serving hundreds of companies and projects, I had enough track record to measure the impact my work had on those organizations, both from a business and a cultural standpoint. "Good design is good business" could have been the motto of my interventions, but unfortunately, IBM's Thomas Watson, Jr. had already come up with this tagline. Design is a field where I felt legitimate, enthusiastic, and excited about using my skills to pursue greater goals and tackle important challenges. This personal connection to design is not unique to me but is a sentiment shared by many designers I have met.

For many of us, design represents a versatile and adaptable tool, unlike more rigid fields such as accounting or hardcore sciences for instance, which often feel confined to specific contexts. Design, on the other hand, offers flexibility and creativity, allowing for its application across various domains and industries. This adaptability makes design an invaluable asset in addressing complex problems, including those related to DEI.

The Collaborative Nature of Design

Designers thrive on the ability to see the world through a different lens. They are trained to observe, empathize, and innovate—qualities that are essential when tackling DEI challenges. By understanding diverse user needs and experiences, designers can create solutions that are inclusive and equitable. This process often involves iterative testing and feedback, ensuring that the final product truly meets the needs of all users, not just a select few.

A key aspect of the design process is vulnerability and creating a safe space with users. Designers must go beyond the surface to truly understand the challenges faced by different communities. This requires humility and the willingness to accept and incorporate feedback. The best designs come from a collaborative effort, working closely with users and communities to refine ideas and solutions continuously.

Simplification and Systemic Challenges

Another principle that guides designers is the pursuit of simplicity. As the saying goes in design, "Perfection is achieved not when there is nothing more to add, but when there is nothing left to take away." This drive to simplify pushes designers to challenge the system and the overall user experience rather than just the product. This approach is particularly effective in addressing DEI issues, as it focuses on removing barriers and creating more accessible and intuitive solutions for everyone.

Restless Testing and Feedback Integration

The iterative nature of design—constantly testing, refining, and iterating—ensures that solutions remain relevant and effective. This relentless pursuit of improvement is coupled with a deep sense of humility. Designers understand that no solution is perfect and that the best results come from continuously seeking and incorporating feedback. This mindset is crucial in creating products and services that truly serve diverse populations.

Diversity of Perspective

Designers also actively seek collaboration and diversity of perspective. By bringing together individuals with different backgrounds, experiences, and viewpoints, the design process becomes richer and more innovative. This diversity is not just beneficial but essential in creating solutions that are inclusive and equitable. It allows for a broader understanding of user needs and leads to more creative and effective solutions.

The principles and habits that define a designer's mindset are increasingly extending beyond the confines of professional practice, influencing personal lives as well. This permeability allows these values to permeate all areas of life, amplifying their reach and impact. As individuals adopt a design-oriented approach, it not only transforms their professional actions but also shapes their interactions within their families, with friends, and in their roles as consumers. This holistic adoption of design thinking fosters a more intentional, empathetic, and solution-oriented approach to everyday challenges, thereby enhancing both personal and collective well-being.

Creating Safe and Inclusive Spaces

In addition to focusing on the practical aspects of design, there is a strong emphasis on creating emotionally safe and inclusive spaces. This involves understanding the cultural and social contexts of the users and designing products and services that resonate on a deeper level. By fostering a sense of belonging and representation, designers can create experiences that are not only functional but also emotionally engaging and supportive.

Design is more than just a profession; it is a powerful tool that can drive meaningful and lasting change. Its versatility, adaptability, and the enthusiasm it inspires make it an ideal field for addressing DEI issues and fostering a more inclusive and equitable society. By leveraging the unique strengths of design, such as collaboration, empathy, simplification, and continuous improvement, we can create solutions that are not only effective but also deeply resonant with the diverse needs of our communities.

In conclusion, the approach inherent to design—marked by vulnerability, collaboration, and an unyielding quest for simplicity and improvement—aligns perfectly with the goals of DEI. By embedding these principles into the design process from the outset, we can create products, services, and environments that are inclusive, equitable, and beneficial for all. This foundational role of design makes it an essential lever for driving meaningful and lasting change in society.

Historical Context and Evolution of Inclusive Design

The concept of Inclusive Design has evolved significantly over the years, influenced by various social, technological, and cultural shifts. Here is a summary of its historical context and evolution.

Early Beginnings: Universal Design

The origins of Inclusive Design can be traced back to the mid-20th century with the emergence of universal design, a concept that sought to create products and environments accessible to all people regardless of their age, abilities, or other personal characteristics. This idea was revolutionary in its departure from reactive accessibility measures, focusing instead on proactive designs that inherently worked for the widest possible range of users.

A key figure in this movement was Ron Mace, an architect and advocate who coined the term *universal design* in the 1970s. Having contracted polio as a child and experienced life in a wheelchair, Mace brought a personal understanding of the challenges faced by people with disabilities. His belief that design should enable dignity and independence for everyone fueled his advocacy and innovations. His contributions crystallized in the creation of the Seven Principles of Universal Design. These principles—covering equitable use, flexibility, intuitive functionality, perceptible information, tolerance for error, low physical effort, and appropriate size and space for approach and use—have since become a guiding framework for architects, urban planners, and product developers.

One example of universal design's success can be found in curb cuts, the small ramps at sidewalk edges originally intended to accommodate wheelchair users. Over time, these features proved invaluable for many others, from parents with strollers to delivery workers and cyclists, demonstrating how inclusive designs can benefit a wide audience. Similarly, automatic doors, initially designed to ease access for people with mobility challenges, have become an everyday convenience, aiding shoppers with full hands, travelers with luggage, and countless others.

Universal design's influence extends beyond physical spaces to everyday products and work environments. Large-print labels and braille packaging, for instance, serve not only those with visual impairments but also older adults with declining vision and even people in dimly lit settings. In workplace design, flexible workspaces featuring adjustable desks, ergonomic chairs, and accessible layouts support employees with diverse needs while enhancing comfort and productivity for all.[12–14]

These examples underscore the ethos of universal design: creating solutions that seamlessly integrate usability and accessibility. Over time, this approach has evolved into Inclusive Design, which shifts the emphasis from broad accessibility to a deeper focus on usability and equity. Although universal design laid the groundwork by demonstrating that inclusive practices improve lives, Inclusive Design builds on this by actively reducing barriers and empowering marginalized groups.

The Disability Rights Movement

The disability rights movement of the 1960s and 1970s marked a watershed moment in the evolution of Inclusive Design principles. Activists during this period relentlessly fought to secure accessibility, equal rights, and the removal of systemic barriers that excluded people with disabilities from full participation in society. Their advocacy was not just about highlighting the obstacles faced by people with disabilities but about demanding substantive changes to laws, infrastructure, and social attitudes to guarantee their dignity and rights.

A landmark achievement of this movement was the enactment of the Americans with Disabilities Act (ADA) in 1990, a transformative piece of legislation in the United States. The ADA did more than establish legal protections—it signaled a societal commitment to inclusivity and equal opportunity. It required public spaces, buildings, transportation, and even digital platforms to be accessible, fundamentally reshaping how designers, architects, and urban planners approached their work.

Before the ADA, much of the built environment was inaccessible to people with mobility impairments. The law mandated retrofits, such as ramps, elevators, and other features that transformed public buildings into accessible spaces. Public transportation systems were overhauled, with innovations like low-floor buses and subway stations equipped with ramps and lifts, enabling independent travel for wheelchair users. The digital realm, too, underwent a revolution, with accessibility standards like the Web Content Accessibility Guidelines (WCAG) emerging to ensure people with visual, auditory, and cognitive impairments could navigate and use online services effectively.

The ripple effects of the ADA extended into product design, fostering the ethos of "designing for all." Products like OXO's Good Grips kitchen utensils, initially developed for people with arthritis, became widely popular because of their ergonomic design, demonstrating that inclusivity can enhance usability for everyone. Similarly, tactile paving at crosswalks and braille signage in elevators became standard features, reflecting a broader commitment to inclusive public spaces.

This period also highlighted the intersection of legal mandates and societal attitudes. The changes spurred by the ADA and similar global efforts were not just about compliance but also about embedding the principle that accessibility is a moral imperative. The disability rights movement reframed the conversation around design, showing that inclusivity benefits everyone and fosters a more equitable society.

These advances set the foundation for the broader concept of Inclusive Design. By normalizing the idea that spaces, products, and services should accommodate the diversity of human experiences, the movement underscored the importance of viewing inclusivity not as a technical afterthought but as an essential element of societal progress. The lessons of this era remain vital today, as they inspire continued efforts to design a world where everyone can participate fully and equally.[15–19]

Design for All in Europe

In the late 20th century, Europe embraced the Design for All movement, a philosophy that paralleled universal design principles but placed a distinct emphasis on participation and inclusivity. Rather than retrofitting existing designs to accommodate marginalized groups, this approach integrated accessibility from the outset, ensuring that products, spaces, and services were inherently usable for everyone—particularly people with disabilities, the elderly, and diverse cultural communities.

A key proponent of this movement was the European Institute for Design and Disability (EIDD),[20] founded in 1993, which pushed for a shift from compliance-driven design to true equity. Instead of treating marginalized groups as passive beneficiaries, the institute advocated for their active involvement in shaping the environments and products that affected their daily lives. This philosophy reshaped urban planning, public policy, and technological innovation across Europe, embedding inclusivity as a fundamental design principle rather than an afterthought.

One of the most visible successes of this movement was the transformation of public spaces in cities like Barcelona and Copenhagen, where inclusive urban infrastructure—tactile paving, ramps, and accessible public transportation—set new global standards. Policymakers also embraced participatory design, with Sweden requiring that people with disabilities be directly involved in shaping public housing and transportation projects to ensure practical usability.

Cultural spaces followed suit, integrating diverse perspectives into their design and programming. A prime example is the Museum of World Culture in Gothenburg, Sweden[21] an institution dedicated to exploring global cultural narratives. Its design prioritizes accessibility, featuring adaptable spaces, multisensory exhibits, and multilingual resources to engage visitors from all backgrounds.

Beyond ethical considerations, Design for All proved to be a compelling business case. By developing products and services that catered to broader demographics, companies expanded their customer base, enhanced user satisfaction, and increased profitability. The movement demonstrated that inclusivity fosters better, more innovative solutions that benefit not only marginalized users but society as a whole. Today, these principles continue to shape global approaches to Equitable Design, proving that accessibility is not just a moral obligation—it's a driver of progress and economic opportunity.

Technological Advancements and Digital Inclusion

The rapid advancement of technology in the late 20th and early 21st centuries brought new dimensions to Inclusive Design. The rise of digital technologies necessitated the creation of accessible websites and software. Thinkers like Tim Berners-Lee, the inventor of the World Wide Web, emphasized the importance of web accessibility. The WCAG, first published in 1999 by the World Wide Web Consortium (W3C), set standards for making web content accessible to people with disabilities.

Here is the WCAG's table of contents from 1999, pioneering and paving the way for inclusive digital practices:[22]

Web Content Accessibility Guidelines:

1. Provide equivalent alternatives to auditory and visual content.
2. Don't rely on color alone.
3. Use markup and style sheets and do so properly.
4. Clarify natural language usage.
5. Create tables that transform gracefully.
6. Ensure that pages featuring new technologies transform gracefully.
7. Ensure user control of time-sensitive content changes.
8. Ensure direct accessibility of embedded user interfaces.
9. Design for device-independence.
10. Use interim solutions.
11. Use W3C technologies and guidelines.
12. Provide context and orientation information.
13. Provide clear navigation mechanisms.
14. Ensure that documents are clear and simple.

Human-Centered Design

The evolution of Inclusive Design also intersects with the principles of human-centered design. This approach, popularized by design firms like IDEO and thinkers such as Don Norman, focuses on designing products with the end user in mind. Often considered the "father of user experience," Norman pioneered the concept of human-centered design, which emphasizes designing with a deep understanding of users' needs, behaviors, and emotions. His work has significantly influenced how design is approached, particularly in creating equitable and inclusive solutions, and his philosophies provide an essential underpinning for many of the ideas explored in this book.

Human-centered design emphasizes empathy, understanding user needs, and iterative testing, all of which align with the goals of Inclusive Design.

Modern Inclusive Design

In recent years, Inclusive Design has gained broader recognition and application across various industries. Tech companies like Apple and Microsoft have made significant strides in developing accessible technology. For example, Apple's VoiceOver and Microsoft's adaptive accessories showcase how Inclusive Design can be integrated into mainstream products.

The historical context and evolution of Inclusive Design demonstrate its deep roots in various social movements, technological advancements, and design philosophies. From universal design and the disability rights movement to modern digital inclusion and human-centered design, the journey of Inclusive Design reflects an ongoing commitment to creating products and environments that are accessible, usable, and beneficial for everyone. This rich history underscores the importance of continuing to innovate and advocate for inclusivity in all aspects of design.

Notes

[1] National Action Network, 2023. On third anniversary of George Floyd's murder, civil rights coalition puts a spotlight on the state of corporate DEI pledges. [Online]. Available at: `https://nationalactionnetwork` `.net/newnews/on-third-anniversary-of-george-floyds-murder-civil-` `rights-coalition-puts-a-spotlight-on-the-state-of-corporate-` `dei-pledges` [Accessed: 16 January 2025].

[2] Diversity Resources, 2023. How DEI has changed 2 years after George Floyd. [Online]. Available at: `https://www.diversityresources.com/` `how-dei-has-changed-2-years-after-george-floyd` [Accessed: 16 January 2025].

[3] CNBC, 2023. Black leaders rate corporate America's DEI progress since George Floyd. [Online]. Available at: `https://www.cnbc.com/2023/06/19/` `black-leaders-rate-corporate-americas-dei-progress-since-george-` `floyd.html` [Accessed: 16 January 2025].

[4] Senior Executive, 2023. Three years after George Floyd's murder: Where is DEI now, and what have companies learned? [Online]. Available at: `https://seniorexecutive.com/three-years-after-george-floyds-murder-where-is-dei-now-and-what-have-companies-learned` [Accessed: 16 January 2025].

[5] NBC News, 2023. Diversity roles disappear three years after George Floyd protests inspired them. [Online]. Available at: `https://www.nbcnews.com/news/nbcblk/diversity-roles-disappear-three-years-george-floyd-protests-inspired-rcna72026` [Accessed: 16 January 2025].

[6] McKinsey & Company, 2023. Most diverse companies now more likely than ever to outperform financially. [Online]. Available at: `https://www.mckinsey.com/featured-insights/sustainable-inclusive-growth/chart-of-the-day/most-diverse-companies-now-more-likely-than-ever-to-outperform-financially` [Accessed: 16 January 2025].

[7] Centers for Disease Control and Prevention (CDC), 2023. Vision loss prevalence. [Online]. Available at: `https://www.cdc.gov/vision-health-data/prevalence-estimates/vision-loss-prevalence.html` [Accessed: 16 January 2025].

[8] National Institutes of Health (NIH), 2016. Visual impairment and blindness cases in the U.S. expected to double by 2050. [Online]. Available at: `https://www.nih.gov/news-events/news-releases/visual-impairment-blindness-cases-us-expected-double-2050` [Accessed: 16 January 2025].

[9] United States Census Bureau, 2015. Projections of the size and composition of the U.S. population: 2014 to 2060. [Online PDF]. Available at: `https://www.census.gov/content/dam/Census/library/publications/2015/demo/p25-1143.pdf` [Accessed: 16 January 2025].

[10] CNN, 2023. Dolce & Gabbana partners with Karen Mok for new collection in China. [Online]. Available at: `https://www.cnn.com/style/article/dolce-gabbana-karen-mok-china/index.html` [Accessed: 16 January 2025].

[11] Hoyer, D., Bennett, J. S., Reddish, J., et al., 2022. Navigating polycrisis: long-run socio-cultural factors shape response to changing climate. *Phil. Trans. R. Soc. B*, 378, 20220402 [Online]. Available at: `https://royalsocietypublishing.org/doi/10.1098/rstb.2022.0402` [Accessed: 16 January 2025].

[12] Mace, R. L., Hardie, G. J. & Place, J. P., 1991. Accessible environments: Toward universal design. In: W. E. Preiser, J. C. Vischer & E. T. White, eds. *Design intervention: Toward a more humane architecture*. Van Nostrand Reinhold.

[13] Steinfeld, E. & Maisel, J., 2012. *Universal design: Creating inclusive environments*. John Wiley & Sons.

[14] Clarkson, P. J. & Coleman, R., 2015. History of inclusive design in the UK. *Applied Ergonomics*, 46, 235–247.

[15] Hamraie, A., 2017. *Building access: Universal design and the politics of disability.* University of Minnesota Press.

[16] Pelka, F., 2012. *What we have done: An oral history of the disability rights movement.* University of Massachusetts Press.

[17] Williamson, B., 2019. *Accessible America: A history of disability and design.* NYU Press.

[18] Fleischer, D. Z. & Zames, F., 2011. *The disability rights movement: From charity to confrontation.* Temple University Press.

[19] Pullin, G., 2009. *Design meets disability.* MIT Press.

[20] European Institute for Design and Disability (EIDD), 1993. `https://dfaeurope.eu`.

[21] A prime example is the Museum of World Culture in Gothenburg, Sweden, `https://www.varldskulturmuseerna.se/en/#`.

[22] World Wide Web Consortium (W3C), 1999. Web content accessibility guidelines 1.0 (WAI-WEBCONTENT). [Online]. Available at: `https://www.w3.org/TR/WAI-WEBCONTENT` [Accessed: 16 January 2025].

The Inclusive Design Mindset

The Inclusive Design mindset is more than a professional approach—it is a lens for viewing the world through empathy, equity, and deliberate action. It challenges us to rethink how we create, communicate, and connect, ensuring that our choices promote inclusion and accessibility for all.

My journey as an educator at institutions like UC Berkeley and MIT, combined with my work as a consultant for leading global corporations, has shaped this perspective. In classrooms, I've guided aspiring leaders through the intricacies of design, technology, and societal impact. In boardrooms, I've collaborated with tech innovators and consumer goods giants to weave inclusivity into their products and services. These experiences have not only informed my philosophy but also provided me with the tools to create practical frameworks that resonate with designers and decision-makers alike.

Through this dual lens of academic rigor and real-world application, I've seen how Inclusive Design transforms outcomes. It illuminates blind spots, dismantles systemic barriers, and empowers people and organizations to make impactful changes. Whether you are a corporate leader shaping the future of innovation or an individual exploring your role in fostering equity, this mindset offers a powerful framework for action.

It is critical to recognize that design is often misunderstood as a superficial exercise in aesthetics. In this book, *Design* encompasses the entire creative process—a problem-solving methodology that shapes systems, decisions, and experiences. It is this deeper understanding of design that equips us to tackle societal challenges with precision and purpose. The following sections delve into cultivating this mindset and applying an inclusive lens to everyday decisions, empowering you to drive meaningful change in your personal and professional spheres.

This chapter explores the multifaceted nature of the Inclusive Design mindset, breaking it down into actionable strategies and insights. We begin by examining how to develop an inclusive lens in daily life, understanding the role of individuals as both citizens and creators of change. From there, we delve into mastering core diversity, equity, and inclusion (DEI) concepts, embracing the full spectrum of diversity, and addressing biases through thoughtful design. Finally, the chapter highlights the importance of vulnerability in fostering empathy and trust, as well as the need to redefine societal narratives to create a more equitable and inclusive future. By integrating these principles into our decisions and actions, we can reshape systems and spaces to empower every individual.

Developing an Inclusive Lens in Everyday Life

As we delve into adopting an inclusive lens in our daily lives, it's crucial to recognize that our choices do not exist in isolation—they influence the broader systems, norms, and structures we inhabit. Design, in this context, goes beyond shaping products and services; it impacts the societal frameworks that govern how we live, work, and interact. This chapter extends the principles of Inclusive Design from professional domains into the sphere of citizenship, emphasizing how our roles as consumers, advocates, and community participants can drive equity and inclusion.

This book is crafted for leaders seeking tangible strategies to integrate inclusion into their organizations, for designers eager to amplify the societal impact of their work, and for active citizens who want to foster a more equitable world. Whether your setting is a corporate boardroom, a design studio, or a community space, this framework equips you to act with purpose and intention, demonstrating that meaningful change begins with everyday decisions.

My journey advising global companies and educating leaders at prestigious institutions like UC Berkeley and MIT has underscored the importance of bridging personal responsibility with collective action. Inclusion is not merely a series of individual efforts; it requires a cohesive understanding of how our actions interconnect. As we transition from examining personal practices to exploring their societal implications, we'll investigate how each of us can serve as catalysts for positive change in the world around us.

Understanding Your Role as a Citizen and a Consumer

The role of a citizen is one of profound responsibility and opportunity. As citizens, we inherit the rights and freedoms established by our predecessors, but we also inherit the duty to expand and protect these rights for future generations. My work as an innovation leader and educator has reinforced my belief that societal change begins with the choices we make—whether in boardrooms, classrooms, or local communities.

In the next section, we'll delve deeper into the intersection of civic responsibility and Inclusive Design. At its heart, the fight for justice and liberty is a fight for equity. By drawing on historical examples and personal experiences, we'll examine how the principles of Inclusive Design can help bridge the gaps in our society, ensuring that justice and liberty are not privileges reserved for a few but rights upheld for all.

Justice and Liberty for All

As citizens, we are granted rights and responsibilities that form the bedrock of our society. These rights and responsibilities vary from country to country, but in the United States, two of the most essential rights are liberty and justice. These principles are enshrined in the nation's founding documents and are fundamental to the American way of life.

However, with these rights comes a corresponding responsibility. Although we are entitled to enjoy liberty and justice, we also bear the duty to protect and ensure that these rights are extended to all citizens. This is not just a personal obligation but a societal one, reflecting the core values of community and solidarity. The Inclusive Design mindset is deeply rooted in this principle. Whether you feel deprived of these fundamental rights or recognize that some of your fellow citizens are not fully receiving them, you have both the power and the responsibility to think, act, and contribute to the realization of these essential elements of our society.

Expanding on this, the responsibility to uphold liberty and justice for all isn't merely a passive duty; it calls for active participation. It means challenging the systems and practices that perpetuate inequality, advocating for policies that promote inclusivity, and supporting initiatives that ensure everyone—regardless of race, gender, ability, or socioeconomic status—has equal access to opportunities. It's about recognizing that the benefits of these rights are not fully realized until they are universally applied.

Furthermore, embracing this responsibility can lead to meaningful change, not just for others but for oneself. By actively participating in the promotion of liberty and justice, we strengthen our communities, build trust among citizens, and create a more equitable society. This proactive stance is a cornerstone of

Inclusive Design, which seeks to create environments, products, and services that cater to the diverse needs of all people.

Nelson Mandela famously said, "For to be free is not merely to cast off one's chains, but to live in a way that respects and enhances the freedom of others." This powerful statement encapsulates the notion that true freedom is intertwined with the freedom of others. Mandela's life and leadership exemplify this belief, as he dedicated himself to dismantling apartheid and building a South Africa where everyone could experience liberty and justice.

This message resonates with me on a personal level. I spent my early years in South Africa, between 1990 and 1995, during the final years of apartheid and the dawn of a new democratic era. These formative years, set against the backdrop of Nelson Mandela's rise to the presidency, were profoundly influential. Witnessing the transformation of a nation from oppression to a more inclusive society left an indelible mark on me. Mandela's leadership and the societal changes he spearheaded are vivid reminders of the power of Inclusive Design and the importance of ensuring that every citizen enjoys the rights of liberty and justice.

What makes Mandela's approach even more remarkable is that once he attained power, he chose a path of reconciliation rather than retribution. Despite having endured decades of imprisonment and witnessing the harsh realities of apartheid, Mandela understood that the future of South Africa depended on creating a society where former oppressors and victims could coexist and work together. He recognized that a stable and equitable society could only be achieved by avoiding cycles of vengeance and fostering a spirit of unity. Mandela's commitment to reconciliation is a powerful testament to the inclusive mindset—one that seeks to build bridges and create systems where all individuals, regardless of their past, can contribute to a more just society.

Mandela's words remind us that our own freedom is intrinsically linked to the freedom of others. His leadership during South Africa's transition serves as a powerful reminder that the fight for liberty and justice is ongoing and that each of us has a role to play in ensuring that these rights are upheld for all. The Inclusive Design mindset, rooted in these principles, is not just a professional approach but a moral and civic duty to ensure that the liberties we cherish are shared and protected for all members of society.

Vote with Your Wallet

Next, let's revisit the concept of "voting with your feet" or "voting with your wallet," which was introduced earlier in this book. This idea isn't just theoretical—it's a powerful tool that can shape the world around us. A few years ago, during my work with the Equitable Design Lab program, I collaborated with a forward-thinking startup that sought to classify businesses based on their stances and actions on various social issues. They were using an AI model—a groundbreaking approach at the time—to analyze and sort companies according to specific traits. For example, they examined how Fortune 500 companies positioned themselves on critical issues like abortion rights by crunching data from press releases, donations, and the causes these companies supported. From this data, the startup could create detailed profiles of each company, highlighting where it stood on key topics that reflect our values.

The implications of this technology for consumers can be very immediate and powerful. It allows individuals to compare their own values with those of companies and adjust their purchasing decisions accordingly. In many instances, major companies are actively supporting causes that may not align with your personal beliefs. This tool will empower consumers to make more informed choices, enabling them to align their values with their everyday decisions, whether it's choosing which products to buy, which brands to support publicly, or even which company's actions to contribute to.

With these services as you shop, you're considering not just the price tag but also the political and ethical orientation of the brand. We will be able to check our accounts not just for financial metrics but also for a company's carbon footprint, its contributions to rebuilding communities in need, and much more. This is where your role as a citizen becomes crucial. We are all change-makers, ambassadors, influencers, investors, and designers in our own right. By advocating, spending, sharing, and making decisions in our daily lives, we shape the world around us.

Your power as a consumer extends far beyond the individual products you purchase. Every decision you make in the marketplace sends a message, not just to the companies you support but also to the people around you. When you choose to buy from brands that prioritize sustainability, equity, and ethical practices, you're not only making a personal statement—you're setting an example for others to follow. This is where the concept of "voting with your wallet" becomes truly impactful.

Let's consider the ripple effect that can occur when your responsible consumption habits influence your friends, family, and coworkers and even your broader community. As you educate and encourage those around you to consider the social and environmental impact of their purchases, you help create a growing movement of informed consumers. This collective action has the potential to transform the marketplace, shifting the power dynamics so that engaged consumers become a loud majority rather than a silent one. When this happens, large organizations and producers can no longer ignore the demand for inclusivity, sustainability, and fairness.

This phenomenon perfectly illustrates the mindset of Inclusive Design—understanding that every intentional action, no matter how small it may seem, contributes to a larger, positive change. By being mindful of the products and services you support and by spreading this mindset to others, you play a crucial role in shaping a more equitable and inclusive society.

Mastering Essential DEI Concepts

Over the years, the concepts of inclusivity and equality have been brought together under the expansive umbrella of the DEI acronym—diversity, equity, and inclusion. Just as the term *sustainability* can evoke a wide array of topics related to climate, environment, and biodiversity, DEI serves as a unifying framework that encompasses various aspects of social justice and organizational change.

This consolidation has significant advantages, as it fosters a common understanding and provides a shared platform for different actions and initiatives. It raises awareness of the challenges and opportunities associated with DEI, helping organizations and individuals alike to approach these issues holistically.

However, it's crucial to recognize that diversity, equity, and inclusion are distinct concepts, each with its own meaning and implications. As our understanding evolves, additional letters and concepts are often added to the acronym, reflecting the growing complexity and nuance of this field.

There is extensive literature available on each of these concepts, and I encourage readers to explore these resources, which are frequently updated to reflect the latest thinking and best practices. In the following sections, I'll provide a brief overview of these differences to establish a common glossary and to illustrate the broad scope of what DEI can encompass.

- ▪ **D - Diversity**: Diversity refers to the presence of differences within a given setting, encompassing various dimensions such as race, gender, age, ethnicity, ability, sexual orientation, and socioeconomic background. It's about recognizing and valuing the unique attributes that each individual brings to a group or organization.

- ▪ **E - Equity**: Equity involves creating fair opportunities and access for all individuals by recognizing and addressing imbalances and barriers that prevent equal participation. Unlike equality, which treats everyone the same, equity ensures that individuals receive the support they need based on their unique circumstances.

- ▪ **I - Inclusion**: Inclusion is the practice of creating an environment where all individuals feel welcomed, respected, and valued, regardless of their differences. It's about ensuring that diverse individuals have a voice, are involved in decision-making processes, and can fully participate in the community or organization.

- ▪ **B - Belonging**: This addition emphasizes creating an environment where all individuals feel accepted, valued, and included. Belonging goes beyond diversity and inclusion, ensuring that everyone feels they are an integral part of the community or organization.

- **A - Accessibility**: This highlights the importance of ensuring that all environments, products, and services are accessible to people with disabilities. Accessibility ensures that everyone, regardless of physical or cognitive abilities, can fully participate.

- **J - Justice**: This focuses on addressing systemic inequities and advocating for fair treatment and opportunities for all individuals. Justice involves actively dismantling barriers that have historically marginalized certain groups.

- **I - Intersectionality**: This addition underscores the complexity of identities and how different aspects of a person's identity (such as race, gender, sexuality, and class) intersect and impact their experiences and opportunities.

This list is not exhaustive and should be regularly updated to reflect new insights, emerging challenges, and evolving societal values, ensuring that our approach to inclusivity remains dynamic and forward-thinking.

The complexity and beauty of DEI lie in its rich, multifaceted nature, encompassing a vast array of topics, actions, and impacts that reach into every aspect of society and organizational life. This complexity can be challenging because mastering DEI requires not only a deep understanding of each element—diversity, equity, and inclusion—but also a holistic, systemic approach to effectively integrate these principles into everyday practices. It demands a high level of literacy in both the theoretical and practical applications of DEI, as well as the ability to see the connections between different issues and understand how they interact within the broader societal framework.

However, this very complexity is also what makes DEI so profoundly beautiful. It offers a rich tapestry of opportunities for creativity and self-expression, allowing individuals and organizations to craft unique strategies that resonate with their specific contexts and values. Rather than being perceived as a compliance burden, DEI should be viewed as a dynamic and evolving platform that invites innovation and fosters a deeper connection to the human experience. It is an arena where diverse perspectives can come together to create more inclusive and equitable environments, ultimately enriching the fabric of our communities and workplaces. The nuances within DEI are not just challenges to overcome; they are opportunities to make meaningful, lasting contributions to a more just and equitable world.

It's essential to recognize that DEI initiatives can sometimes feel overwhelming, especially for those who are new to the conversation or feel distant from the issues at hand.

Throughout my career, I've had many enriching and thought-provoking discussions with individuals in large organizations. These conversations often revolved around how we could integrate our Equitable Design framework within existing corporate structures. Naturally, these discussions led us to engage with people in positions of influence—managers, decision-makers, and executives who had the power to champion and fund these efforts. However, even in 2024, we frequently encounter a familiar scenario: most of the individuals in these roles tend to be older, white men with privileged backgrounds and high levels of education. This demographic reality often complicates the conversation.

There is often genuine enthusiasm for the concepts we present. I vividly recall one senior manager at a major American corporation who was clearly excited about the potential of making the company's services more inclusive for older adults from underserved communities. Yet despite his enthusiasm, it became evident that he required a significant amount of internal validation before moving forward.

After three productive meetings with him alone, he decided to expand the discussion by inviting five members of his extended team, most of whom were women of color. Although the dialogue was engaging, there was an unmistakable discomfort in the room. The manager, careful not to overstep, appeared tentative in his approach, and some of the invited team members seemed uncertain about their roles or felt pressured to contribute expert-level insights to the conversation. Ultimately, the project stalled as the decision-making process was continually deferred to others.

This experience highlights several recurring themes in our DEI-related discussions. One of the most significant barriers is imposter syndrome and the fear of failure or backlash. Many people, particularly those who don't feel directly connected to DEI issues, struggle to champion these initiatives because they fear they lack the legitimacy to do so. Additionally, there's often a palpable concern about being perceived as engaging in "diversity washing"—the fear that their efforts could be seen as insincere or as an attempt to exploit a cause for image enhancement.

This example underscores the importance of building DEI literacy at all levels of an organization. Understanding the nuances and challenges of DEI work is crucial for fostering genuine and effective initiatives that resonate throughout the company and beyond.

Embracing Diversity's Full Spectrum

Diversity encompasses a wide range of human differences and can be categorized into several main types. These categories help organizations and societies understand and address the various aspects of diversity to foster more inclusive environments.

- ▪ **Demographic diversity**
    - ▪ **Race and ethnicity**: Differences based on physical characteristics and cultural backgrounds
    - ▪ **Gender and sex**: Diversity in gender identities and expressions, and biological differences

- **Age**: Inclusion of various age groups, from young to elderly
- **Sexual orientation**: Diversity in sexual preferences and identities
- **Experiential diversity**
  - **Socioeconomic status**: Variations in income, education, and occupation
  - **Education**: Different educational backgrounds and levels of attainment
  - **Geographical location**: Differences in the environment where individuals were raised or currently reside
  - **Life experiences**: Diverse personal and professional experiences that shape an individual's perspective
- **Cognitive diversity**
  - **Thought styles**: Different ways of thinking, processing information, and problem-solving
  - **Personality traits**: Diversity in personality types, including introversion, extroversion, and various behavioral traits
  - **Abilities and disabilities**: Range of physical, mental, and emotional abilities
  - **Neurotype**: Differences in cognitive functioning and neurological wiring, including conditions such as ADHD, autism, and dyslexia
- **Cultural diversity**
  - **Language**: Inclusion of different languages spoken and understood
  - **Religion and belief systems**: Diversity in religious affiliations, spiritual beliefs, and cultural practices
  - **Customs and traditions**: Varied cultural norms, rituals, and practices
- **Health diversity**
  - **Mental health**: Differences in mental health status, including conditions like depression, anxiety, and PTSD
  - **Physical health**: Variation in physical health, including chronic illnesses or conditions like diabetes and heart disease
  - **Nutritional needs**: Differences in dietary restrictions or preferences influenced by health conditions, cultural practices, or personal choices
- **Generational diversity**
  - **Baby Boomers**: Individuals born approximately between 1946 and 1964, often characterized by their experience and strong work ethic
  - **Generation X**: Born between 1965 and 1980, this group values independence, flexibility, and work-life balance

- **Millennials (Generation Y)**: Born between 1981 and 1996, Millennials are known for their tech-savviness and desire for meaningful work
- **Generation Z**: Born after 1997, Generation Z values diversity, inclusivity, and social responsibility

- **Geographic diversity**
  - **Urban vs. rural**: Differences in perspectives based on whether individuals are from urban, suburban, or rural areas
  - **Regional diversity**: Variations in cultural practices, dialects, and social norms based on specific regions
  - **Global diversity**: Inclusion of individuals from different countries, bringing a variety of cultural, linguistic, and economic perspectives

- **Organizational diversity**
  - **Job function**: Diversity in roles and responsibilities within an organization
  - **Tenure**: Variations in the length of time individuals have been with an organization
  - **Employment status**: Full-time, part-time, freelance, or temporary employment statuses
  - **Work experience**: Different levels of experience and expertise in a field or industry

Understanding these diverse categories is crucial in the mindset of Equitable and Inclusive Design because they illuminate the broad spectrum of human experience and highlight the importance of designing with a truly open mind. When we recognize and appreciate the full range of diversity—whether it's demographic, cognitive, cultural, or otherwise—we are better equipped to create products, services, and environments that are genuinely inclusive. This means acknowledging that every person brings a unique perspective shaped by their background, experiences, and identity.

Incorporating this mindset requires us to challenge our assumptions and accept that our own perspective may be limited. We all view the world through a lens shaped by our individual experiences, and this lens can sometimes obscure problems or needs that others might see clearly. For example, someone who has never experienced a physical disability might overlook barriers that make a product inaccessible to those who do have a disability. Similarly, a person from one cultural background might not fully understand the needs or preferences of someone from a different culture.

By embracing a broad understanding of diversity, designers and decision-makers can avoid the pitfalls of a one-size-fits-all approach and instead create solutions that consider the varied needs of all users. This approach not only leads to more effective and Equitable Designs but also fosters innovation by encouraging the inclusion of diverse perspectives and ideas. In essence, recognizing the full scope of diversity pushes us to think more critically, listen more carefully, and design more thoughtfully, ensuring that the outcomes truly serve everyone, not just the majority or the most visible groups.

Beyond the current set of diversity criteria, it's important to recognize that new aspects of diversity will continue to emerge in the future. The inequalities we face today are not always the same as those of the past. This is partly because societies tend to address and mitigate certain inequalities over time: if not completely erasing them, then at least reducing their impact. Historical examples include the abolition of slavery in the United States and the strides made toward gender equality in voting rights. Additionally, advancements in public health, such as the eradication of smallpox and the significant reduction of polio cases, highlight how some issues can be addressed effectively.

Another reason for the emergence of new aspects of diversity is the increasing awareness brought about by scientific advancements, social advocacy, and the evolution of our societal values and lifestyles. As our understanding deepens, we often uncover new inequalities or areas that require attention. These are often referred to as *new visibilities*.

A prime example of this is the growing recognition of neurodivergence and mental health-related categories. These areas are gaining prominence as diagnostic techniques improve, allowing us to identify a larger segment of the population affected by these conditions. Additionally, lifestyle changes and societal pressures are contributing to an increase in the number of people who fall into these categories.

This is where the Inclusive Design mindset begins. It's not about focusing solely on the most prominent or well-known diversities or minorities. Instead, it's about approaching design with objectivity and neutrality without prioritizing one form of diversity over another. The goal isn't to rank diversities in order of importance but to recognize where action is needed and where it can make the most impact. Prioritization may be based on urgency, the size of the affected population, or the depth of the inequality, but the overarching aim is to act where it matters most.

Transcending Politics: Equality and Inclusive Design

At the beginning of each semester at UC Berkeley, I like to start my classes with a thought-provoking exercise that gets students to reflect on their perceptions of equality and wealth distribution. It's a simple game, but it never fails to spark deep conversations and a few "aha" moments.

I begin by asking my students to write down the top five jobs that UC Berkeley graduates typically aspire to after graduation. The usual suspects almost always appear on their lists—finance traders, coders, data scientists, consultants, and product managers. These are the roles often associated with prestige, high salaries, and the pinnacle of professional success in our society.

Once they've completed this first task, I ask them to set that list aside and forget about it for a moment. Then, I prompt them to consider something entirely different: I ask them to write down the top five jobs they believe are the most essential to the functioning and well being of our society. This is where the lists start to diverge significantly. Almost invariably, the second list includes doctors, teachers, social workers, firefighters, and—although they often forget at first—farmers. It's a moment of realization when they notice that these two lists have almost nothing in common.

This exercise reveals a striking paradox: the best-trained talents from one of the most prestigious institutions in the country are often funneled into roles that are not necessarily the most essential to the basic functioning of society. Meanwhile, the professions that are most critical to our daily lives and societal well-being—those on the second list—are frequently filled by individuals from less prestigious educational backgrounds.

Next, I challenge the students to rank these jobs from the highest-paying to the lowest-paying. Unsurprisingly, the essential jobs—except for doctors—tend to cluster at the bottom of the income scale. This stark contrast underscores a troubling reality: our society does not reward the most necessary professions with the financial compensation they deserve.

The conversation naturally leads to a broader discussion about equality and wealth distribution. I often quiz my students on global wealth distribution, asking them to compare the wealth of the richest 1% with that of the bottom 50% of the world's population. I also ask them to estimate the order of magnitude of this disparity. Their estimates are almost always significantly off—they generally assume that the distribution is much more equitable than it actually is. According to Oxfam International, as of July 2024, the richest 1% have accumulated $42 trillion in new wealth over the past decade—nearly 34 times more than the entire bottom 50% of the world's population.[1]

Equality in society is not merely a political aspiration; it is a fundamental principle that underpins the stability and prosperity of any community. The concept of equality transcends political ideologies, serving as a core value that promotes fairness, opportunity, and social cohesion. To illustrate the universal importance of equality, we can look at findings from the study "Building a Better America—One Wealth Quintile at a Time",[2] which reveals insights into how Americans across the political spectrum perceive wealth distribution and advocate for a more equitable society.

Understanding Equality Through Wealth Distribution

The study "Building a Better America—One Wealth Quintile at a Time," conducted by Michael I. Norton and Dan Ariely, highlights a significant consensus among Americans regarding the ideal distribution of wealth. The research found that irrespective of political affiliations, most people envision a more equitable distribution of wealth than currently exists. This finding underscores a shared belief in the importance of reducing economic disparities to create a fairer society. The following are the key findings of the study:

- **Perception vs. reality**: The study revealed a stark contrast between the perceived and actual distribution of wealth in the United States. Participants, regardless of their political leanings, significantly underestimated the extent of wealth inequality. When informed about the actual disparities, there was widespread agreement on the need for a more balanced distribution.

- **Ideal distribution**: Across the board, Americans favored a wealth distribution where the top wealth quintile held less wealth than they currently do and the bottom quintiles held more. This ideal scenario reflects a desire for a society where opportunities and resources are more evenly distributed.

- **Consensus for change**: The study demonstrated that the majority of participants, regardless of their political views, supported policies and practices that would lead to greater equality. This includes support for progressive taxation, improved access to education, and healthcare reform, which are seen as pathways to a more equitable society.

Inclusive and Equitable Design Beyond Politics

The principles of inclusive and Equitable Design align with the broader societal aspiration for equality. Inclusive Design seeks to create products, services, and environments that are accessible and beneficial to all, regardless of their background or circumstances. This approach inherently addresses disparities by considering the diverse needs of the population:

- **Enhancing accessibility**: Inclusive Design ensures that everyone, including those with disabilities, can access and use products and services. This promotes equality by removing barriers and enabling full participation in society.

- **Promoting representation**: By involving a diverse range of perspectives in the design process, Inclusive Design fosters representation and ensures that solutions are relevant and effective for all segments of the population. This approach helps to address systemic biases and create more equitable outcomes.

- **Building social cohesion**: Inclusive Design contributes to social cohesion by creating shared experiences and spaces that bring people together. When products and environments are designed with everyone in mind, they promote interaction, understanding, and unity.

A Universal Aspiration and a Strategic Imperative

The pursuit of equality is a shared human aspiration transcending political, cultural, and geographic divides. The findings from the study "Building a Better America—One Wealth Quintile at a Time" underscore a widespread desire for a fairer distribution of wealth and resources, a sentiment echoed across diverse segments of society. The principles of Inclusive Design align seamlessly with this universal yearning, fostering accessibility, representation, and social cohesion. At its core, Inclusive Design is a practical response to inequality, offering actionable pathways to build systems that are fairer and more reflective of diverse needs.

Yet it is important to recognize that the dialogue around DEI is far from seamless or universally accepted. Public figures like Robin DiAngelo and Ibram X. Kendi have brought essential issues of race, privilege, and systemic inequality to the forefront of cultural conversations. Their work has been transformative for many, sparking critical self-reflection and institutional change. However, it has also faced significant resistance and, at times, public ridicule. These reactions reflect broader discomfort with change, particularly when it challenges long-held beliefs or entrenched systems of power.

Rather than ignoring these critiques, we must confront them thoughtfully, understanding that meaningful progress often provokes discomfort. Resistance to diversity advocacy frequently stems from fear, misunderstanding, or frustration with poorly implemented initiatives that feel performative or disconnected from tangible results. For some, diversity efforts are seen as a compliance exercise rather than a genuine opportunity for growth and innovation. Reframing this narrative is essential.

Diversity is not just a moral or ethical imperative—it is a strategic advantage. Organizations that embrace diversity consistently demonstrate better decision-making, increased innovation, and stronger financial performance. For leaders, the question should not be whether to prioritize inclusivity but how to integrate it effectively into their strategies to drive meaningful impact.

This book offers practical guidance to bridge divides and foster collaboration, focusing on solutions that resonate with both advocates and skeptics. The goal is not to impose a singular vision of inclusion but to create shared ownership of its benefits. Later chapters will illustrate how Inclusive Design principles can transform organizations, communities, and markets, highlighting examples of success and lessons learned along the way.

In conclusion, Inclusive Design transcends politics and ideology. It is grounded in the fundamental human values of fairness and equality, offering a unifying framework for building a more equitable society. By embracing these principles, we can foster environments where opportunity and belonging are accessible to all. This shared commitment to inclusivity not only drives societal progress but also strengthens the foundations of innovation, collaboration, and collective growth.

Overcoming Bias Through Design

A fundamental aspect of Inclusive Design is the ability to accept and acknowledge the bias that guides our perspective on life. We are looking at situations that are influenced by our education, experience, and sensitivity, and this will mechanically trigger some bias.

This is known to be very helpful in situations where we are able to jump straight to conclusions instead of looking at a known problem with fresh eyes all the time. Yet this definitely contributes to perpetuating stereotypes and judgments, which cause exclusion and inequalities. It is impossible to operate without any bias, but recognizing them and reducing their negative impact is a great thing to do.

Biases are definitely important in a design process, and Inclusive Design aims at mitigating the risks associated with them.

The Implicit Bias in Generative AI: A Personal Exploration

One morning in late June, as I was sipping my coffee and scrolling through my LinkedIn feed, a particular post caught my attention. It was from a professional contact who was raising a critical issue about Generative AI tools and their inherent biases. The post described an intriguing experiment that immediately piqued my curiosity. The individual had prompted a well-known Generative AI image tool to create two images: one of a literate family and one of an illiterate family. The results were displayed in a visually striking post.

The images were aesthetically pleasing, with bright colors and smiling faces radiating happiness. Yet a stark and unsettling difference was apparent. The literate family was depicted as white, their environment and attire suggesting affluence and comfort. In contrast, the illiterate family had darker skin tones, with cultural markers such as a mustache on the man and a red bindi on the woman's forehead, indicating they were likely from India.

The contrast was not just in appearance but also in the subtle messages conveyed by the AI-generated images. The literate family, basking in bright hues, seemed to embody success and education. Meanwhile, the illiterate family, although equally happy in their demeanor, was portrayed with visual cues that could be interpreted as indicative of lower socioeconomic status. This visual juxtaposition raised questions about the biases encoded within the AI algorithms and the data they were trained on.

Intrigued and somewhat unsettled by this demonstration, I decided to conduct a similar experiment using my preferred Generative AI software. I crafted this prompt: "Can you create two images in the same style: close-ups of one literate family and one illiterate family?" The AI swiftly generated the images, as shown in Figure 2.1 and Figure 2.2.

Figure 2.1: First iteration of the AI rendering after being prompted to create an image of a literate family

Figure 2.2: First iteration of the AI rendering after being prompted to create an image of an illiterate family

The AI tool added this comment: "Here are the close-up images of both families. The first image depicts a well-educated family, and the second one shows a family in a modest living room. If you need any further modifications or details, please let me know!"

The first set of images depicted both families with striking detail. The literate family appeared European or American, characterized by modern attire and a sophisticated backdrop. In contrast, the illiterate family, although also white, had cultural and stylistic cues that hinted at different demographics, possibly suggesting Eastern European or Middle Eastern origins. Their clothing and modest living environment subtly implied a lower economic status.

As I examined the images, I noticed additional cultural markers: the man in the illiterate family had a long beard, and the woman wore a hijab. These features led to assumptions about their background and status. A long beard on the man might be associated with traditional or conservative values, often linked to Middle Eastern or South Asian cultures. The woman's hijab, a headscarf worn by some Muslim women, could indicate a religious and cultural adherence to Islam. These assumptions, whether accurate or not, influence perceptions and reinforce stereotypes about education and socioeconomic status.

I regenerated the prompt to explore the consistency of the AI's output. In the second iteration, as shown in Figure 2.3, the literate family remained extremely consistent with the first image generated, featuring the same hairstyles, clothing, and extremely similar faces. They were consistently portrayed as European or American. Meanwhile, as shown in Figure 2.4, the illiterate family's representation varied more widely, leaning toward Asian demographics. This variability underscored the AI's struggle to maintain a consistent narrative for the illiterate family.

By the third iteration, an interesting shift occurred. For the first time, as shown in Figure 2.6, the illiterate family closely resembled the literate one (see Figure 2.5) in terms of racial demographics, suggesting some randomness or perhaps a broader range in the training data. However, the fourth iteration reverted to the initial pattern, with the literate family depicted as affluent and Western and the illiterate family once again portrayed with cultural markers indicating an Eastern or Asian background.

This experiment highlighted a critical issue: the implicit biases embedded within AI models. These biases are not just theoretical; they manifest in the outputs these models generate, influencing perceptions and reinforcing stereotypes. The consistency of the literate family's depiction as white, compared to the variability and ethnic association of the illiterate family, underscores the need for greater scrutiny and diversity in training data.

Generative AI tools hold immense potential, but they also reflect the biases present in their training datasets. As we continue to integrate AI into various aspects of our lives, it's crucial to address these biases. This ensures that the technology we develop and deploy promotes inclusivity rather than perpetuating harmful stereotypes.

Figure 2.3: Second iteration of the AI rendering after being prompted to create an image of a literate family

Figure 2.4: Second iteration of the AI rendering after being prompted to create an image of an illiterate family

Figure 2.5: Third iteration of the AI rendering after being prompted to create an image of a literate family

Figure 2.6: Third iteration of the AI rendering after being prompted to create an image of an illiterate family

It seems that these Generative AI tools have been created to mimic human thinking learning based on our cognitive systems and processes. Given that many of these AI systems are developed in Western countries, they may reflect Western-centric perspectives.

We've seen in many discussions the idea of biased datasets and the fact that AI has been trained on data that includes human biases. These biases, embedded in the training data, can manifest in the AI's outputs, influencing perceptions and reinforcing stereotypes.

The biases observed in AI outputs often stem from the data used to train these models. AI systems learn patterns and make predictions based on vast amounts of data, which can inadvertently contain historical biases and stereotypes present in society. If the training data predominantly represents certain demographics or cultural contexts, the AI will reflect and perpetuate those biases. For instance, if most images labeled as "literate" in the training data depict white families, the AI will learn to associate literacy with whiteness.

Moreover, the annotations and labels applied to training data can introduce biases. Human annotators who label the data bring their own conscious and unconscious biases, which can influence how data is categorized. This process can reinforce existing stereotypes, such as associating specific ethnic or cultural markers with certain socioeconomic statuses.

Although in my case the two families both seem white, the clothes and hairstyles clearly give some indication of demographics and could suggest that literate families are essentially white and probably European or American, whereas the illiterate ones come from Asia or the Middle East.

This experience has its limitations, and by no means does it suggest that the service I used is inherently racist or intentionally designed to cause harm. Additionally, this was not a rigorous experiment, as neither the methodology nor the dataset used can be considered fully representative.

My aim was twofold: first, to assess how the AI was trained and how it responds to prompts that might elicit biased outputs; and second, to provide an example of why we must continually question the tools and systems we rely on. By doing so, we can identify and address potentially harmful biases embedded within them.

I encourage readers to critically examine the systems they depend on in their own lives, exploring how these tools may be influenced by bias and understanding the potential risks they pose. Only through such scrutiny can we work toward building more equitable and trustworthy technologies.

Understanding Bias: A Double-Edged Sword

Biases are inherent in human cognition, playing a crucial role in how we process information and make decisions. They act as mental shortcuts, allowing us to navigate a complex world with efficiency. However, these same biases can lead to undesired effects, particularly in the context of inclusion and equality.

In this section, we will explore the nature of biases, their benefits, and their pitfalls, referencing insights from the book *Thinking, Fast and Slow* by Daniel Kahneman (Farrar, Straus, and Giroux, 2023) and experiments by Dan Ariely. Daniel Kahneman, a Nobel laureate in Economic Sciences, is a pioneer in behavioral economics. His work, particularly with Amos Tversky on Prospect Theory, demonstrated how cognitive biases and heuristics often lead people to make decisions that deviate from traditional economic models of rationality. This research laid the foundation for understanding real-world decision-making.

Dan Ariely, a renowned behavioral economist and a former student of Kahneman, has furthered these concepts, exploring predictable patterns of irrational behavior and their impact on economic and social choices. Behavioral economics, the field they helped shape, combines psychology and economics to reveal how human behavior is influenced by biases, emotions, and social factors, offering profound insights into policy, business, and design.

The Nature and Function of Biases

Biases are cognitive shortcuts that help us process information quickly and efficiently. They arise from our brain's need to conserve energy and make rapid decisions in an environment filled with vast amounts of information. Daniel Kahneman, in his seminal work *Thinking, Fast and Slow*, describes two systems of thinking: System 1, which is fast, automatic, and often subconscious, and System 2, which is slower, more deliberate, and conscious. Biases primarily operate within System 1, allowing us to make quick judgments without the need for extensive analysis.

For instance, when you see a snake, your immediate reaction is to jump back. This is a result of an evolutionary bias toward survival, allowing you to react swiftly to potential danger. Without such biases, our ancestors would have spent too much time deliberating, possibly leading to fatal consequences.

The Benefits of Biases

Biases are not inherently bad; they serve important functions in our daily lives:

- **Efficiency**: Biases enable us to process information rapidly. For example, heuristics, or rules of thumb, help us make quick decisions without extensive deliberation. This is essential in situations that require immediate action.

- **Survival**: Many biases have evolved to enhance our chances of survival. Fear of heights, for example, is a bias that prevents us from taking unnecessary risks that could result in injury or death.

- **Social cohesion**: Certain biases, such as those that favor in-group members, can foster social cohesion and group identity. This has been crucial in the evolution of human societies, enabling cooperation and mutual support.

The Pitfalls of Biases

Although biases can be beneficial, they also have significant drawbacks, especially when it comes to fairness and inclusion. Here are some key biases that can lead to undesired effects:

- **Confirmation bias:** This is the tendency to search for, interpret, and remember information that confirms our preconceptions. It can lead to the reinforcement of stereotypes and the exclusion of differing perspectives. For example, if a hiring manager believes that a certain demographic is less competent, they may unconsciously seek out information that confirms this belief, disregarding evidence to the contrary.

- **Anchoring bias:** This occurs when individuals rely too heavily on an initial piece of information (the "anchor") when making decisions. In negotiations, the first offer can set the tone and influence the outcome, even if it is arbitrary. This can result in unfair advantages or disadvantages.

- **Availability heuristic:** This bias leads people to overestimate the likelihood of events based on their availability in memory. For instance, media coverage of rare but dramatic events (like plane crashes) can lead people to overestimate their frequency, affecting their behavior and decisions.

- **Stereotyping:** This involves overgeneralizing about a group based on limited information. Stereotyping can lead to discrimination and exclusion, as individuals are judged not on their merits but on preconceived notions.

Experiments Highlighting Biases

Dan Ariely has conducted numerous experiments that illustrate the impact of biases on decision-making. In his book *Predictably Irrational* (Harper Perennial, 2016), Ariely demonstrates how irrational our decisions can be due to biases:

- **The decoy effect:** Ariely showed that the presence of a third, less attractive option (the decoy) can influence our choices. For example, when given a choice between a cheaper, lower-quality product and a more expensive, higher-quality product, introducing a third option that is similar but inferior to the higher-quality product makes people more likely to choose the expensive option.

- **The IKEA effect:** This bias occurs when people place a disproportionately high value on products they partially created. Ariely found that individuals were willing to pay more for furniture they assembled themselves compared to identical preassembled furniture, highlighting the bias toward our own efforts and creations.

Biases in Action: The Four-Cards Exercise

Understanding bias is a critical component of my course, and I emphasize this from the very beginning. To introduce the concept, I engage my students in a simple game during one of our first classes. The task is to determine which card to turn over to correctly solve a problem.

This exercise often generates lively discussion and varying responses. Over the years, I've noticed that the majority of the class tends to choose the wrong answer. However, there are usually one or two students who, after some reflection, manage to grasp the underlying challenge and identify the correct solution.

As I narrate this experience, I invite you, the reader, to take part in the same challenge. Imagine yourself in the classroom, faced with the task—will you fall into the common trap, or will you discern the right path?

The game is the Wason Selection Task, also known as the four-cards exercise. Here's how it works:

You are shown four cards, each with a letter on one side and a number on the other:

- Card 1: A
- Card 2: D
- Card 3: 4
- Card 4: 7

You are given a rule: "If a card has a vowel on one side, then it must have an even number on the other side."

Your task is to decide which two cards you need to turn over to test whether this rule is true.

Take a moment to think it through. Which cards would you choose?

Answer:

1. **Card A (vowel):**
   - **Reason:** This card directly involves the rule because it shows a vowel. You need to check if the other side is an even number. If it's not, the rule is broken.
   - **Action:** Flip this card.

2. **Card D (consonant):**
   - **Reason:** The rule only applies to vowels. Because this card shows a consonant, what's on the other side doesn't matter. The rule doesn't say anything about what should happen if the card shows a consonant.
   - **Action:** Do *not* flip this card.

3. **Card 4 (even number):**
   - **Reason:** The rule only tells you what should happen if there's a vowel on one side, not what happens if there's an even number. Knowing what's on the other side of this card does not help you test the rule.
   - **Action:** Do *not* flip this card.

4. **Card 7 (odd number):**
   - **Reason:** If this card has a vowel on the other side, then the rule is violated because vowels must pair with even numbers. Checking this card can potentially disprove the rule.
   - **Action:** Flip this card.

5. **Conclusion:**
   The correct cards to flip are A and 7.
   - If A has an odd number on the back, the rule is false.
   - If 7 has a vowel on the back, the rule is also false.

Most people mistakenly choose A and 4, thinking they need to confirm the rule. However, in logic problems like this, the goal is to *disprove* the rule by finding exceptions, which is why you flip A (to check for an odd number) and 7 (to check for a vowel). This exercise is a classic demonstration of *confirmation bias*—the tendency to seek out information that confirms our existing beliefs while overlooking or ignoring contradictory evidence.

This game illustrates the importance of challenging assumptions and being open to finding disconfirming evidence—an essential mindset for critical thinking, especially in design thinking and problem-solving contexts.

Addressing Biases for Greater Inclusion

Understanding and addressing biases is crucial for fostering inclusion and equality. Here are some strategies to mitigate the negative effects of biases:

- **Awareness and education**: Educating individuals about biases and their effects can help mitigate their impact. Training programs that highlight common biases and their consequences can foster more equitable decision-making.

- **Diverse teams**: Encouraging diversity in teams can reduce the influence of individual biases. Diverse groups bring different perspectives, which can challenge assumptions and lead to more balanced decisions.

- **Structured decision-making**: Implementing structured processes for decision-making can minimize the impact of biases. For example, using standardized criteria for hiring or promotions can reduce the influence of subjective judgments.

- **Blind evaluations**: Removing identifying information from evaluations can help prevent biases based on gender, ethnicity, or other factors. This approach has been shown to increase diversity in fields like music and academia.

Conclusion

Biases are an integral part of human cognition, enabling us to navigate a complex world efficiently. However, they can also lead to undesired effects, particularly in the context of inclusion and equality. By understanding the nature of biases and implementing strategies to mitigate their impact, we can create more inclusive and equitable environments.

As Kahneman and Ariely's works illustrate, biases are a blessing and a curse. They save us time and effort but can also cloud our judgment and perpetuate inequalities. By recognizing and addressing biases, we can harness their benefits while minimizing their drawbacks, fostering a society that values fairness and inclusivity.

Beyond Empathy—The Role of Vulnerability

In a world that often values strength and invulnerability, it can be challenging to embrace a concept like vulnerability. However, vulnerability is not a weakness; it's a strength that can fuel innovation, foster connection, and inspire change. Embracing vulnerability is key for addressing significant challenges and driving meaningful change, and it's a pillar of the Inclusive Design mindset.

The Power of Vulnerability to Accelerate Transitions

Brené Brown, a research professor who has extensively studied courage, vulnerability, shame, and empathy, asserts that vulnerability is the birthplace of love, belonging, joy, courage, empathy, and creativity. This means vulnerability can lead to breakthrough ideas and strong connections with others. Brown writes, "Vulnerability is not winning or losing; it's having the courage to show up and be seen when we have no control over the outcome".[3] This perspective is crucial for those looking to make a significant impact.

As individuals and members of organizations, we constantly need to change, adapt, and evolve. In my work, I've been at the forefront of many changes, or at least their inception phase. Over the past decade, I've counseled leaders and led design workforces leveraging innovation to remain relevant, worked with thousands of students at prestigious universities, and supported them in unfolding their careers. A common thread in all these contexts is the need to accept that there should be a starting point and an ending point—a transition. Recognizing that people are currently in a situation that is not completely satisfactory and that they would rather be in a slightly different place is essential. Whether it's a company facing market challenges, a student challenging the status quo, or a leader aiming for a sustainable and equitable world, they all engage me to help them make this transition.

The fascinating piece is that a big component of success in these endeavors is to admit, acknowledge, and accept this journey from A to B, from now to then. This process requires courage, lucidity, and skills. This three-stage process is what I call vulnerability: admit, acknowledge, and accept.

Admit Admitting means facing a certain situation with self-alignment. The preface of a transition is usually an emotion, a discomfort that one feels. Statements like "I don't feel good about this work I'm doing," "I feel uncomfortable telling friends I am working for this company," and "I feel lost" are premises of a transition. The goal is to transcribe this emotion into something actionable without deteriorating or diminishing the emotion because it will remain the key driver of change.

Once someone recognizes this discomfort and the need for change, they start admitting it. They grow a desire to change, turning a negative emotion into motivation. The admit stage is fulfilled once we align our emotions, desires, and logic. The moment we can articulate an objective, we reach an important milestone in the process.

Acknowledge The second stage of vulnerability involves acknowledging the transition needed. This is crucial in two ways: first, it forces us to "step out of the closet," displaying courage and determination. Second, it creates a movement; we're not alone anymore. Even if we lack direct support from others, acknowledging and sharing our desire for a transition means sharing some of the burden.

For example, in a company, the moment an individual shares with leadership a pressing challenge or something that is not done right, the leaders become aware and, by extension, responsible for the situation. They can't ignore what they know, and their actions or inactions will determine their accountability.

Acknowledging also offers an opportunity to improve the transition statement. By exposing it to more people, we are forced to rephrase, adjust, and correct our wording to fit different audiences. Meanwhile, the feedback received helps create a more cohesive and elaborate phrasing.

Accept The last step of vulnerability is acceptance. By accepting, we assimilate what comes with the need for transition and stop struggling with ourselves. We create a sense of peace and a positive cycle to proceed.

Acceptance turns negative emotions into positive intentions. As a strategic consultant or a teacher, I have witnessed many teams of professionals and students cross the chasm and enter this acceptance phase. From there, they stop arguing about the problem and who should be held accountable, who is to blame, and how they should be punished. Instead, they move to a constructive solution phase, where each team member suggests approaches and ways to solve the problem. We can hear complementary visions and constructive feedback because we've escaped the judgmental phase and embraced a more empathetic perspective. We have freed people's creativity and put down inhibitions.

Now it's just a matter of time and perseverance until that change, that transition, can happen.

Overcoming the Pressure of "What Will People Think?"

One of the biggest barriers to embracing vulnerability is the fear of what others will think. In my research, I've found that people often base important decisions on what they think others will think rather than on their own values or desires. This mental pressure comes from people who are spectators more than actors—parents, partners' friends, former teachers—who won't be directly affected by the decision but still exert influence.

Moreover, the projections made about what these spectators might think are often inaccurate. It's difficult enough to predict someone's actions, but predicting their thoughts and feelings is even harder. This leads to a case where important decisions are made based on assumptions about spectators.

Using vulnerability in the context of important decisions allows the decision-maker to face the actual actors and base decisions on facts rather than assumptions.

Vulnerability in Leadership

A leader's ability to be vulnerable can greatly influence their effectiveness. By admitting they don't have all the answers, leaders can foster an environment of openness and mutual respect. Although we are trained and encouraged to show no weaknesses, we all know that no one, not even the best leaders, can know everything. Not recognizing this and trying to display perfection sets an unsustainable social contract within a team.

A common misconception is that being vulnerable means inviting judgment and leaving oneself defenseless. In reality, recognizing that you might not have the answer means sharing the responsibility for the decision. You are not better qualified than others, so you're asking for their help. Although we can be judged for making a wrong decision under our own judgment, we can't be held solely responsible for a collegial decision where all the cards have been shown.

Beyond this, vulnerability triggers empathy. We are naturally wired to support the weaker (the victim) and to challenge the stronger (the privileged). A leader showing vulnerability will always earn empathy from their audience, leading to better connections, discussions, and problem-solving approaches.

As we create and benefit from this empathetic environment, we make it available to everyone around us. Showing vulnerability can lead to increased creativity, as team members feel safe to express their ideas and take risks, knowing that they won't be judged. Leaders who demonstrate vulnerability can inspire trust and loyalty, creating a stronger, more cohesive team ready to drive change.

Vulnerability and Resilience

Embracing vulnerability also builds resilience, a critical trait for change-makers. In the face of adversity, it is not invulnerability but resilience that enables us to bounce back. By accepting and embracing our vulnerability, we can learn to cope with difficulties, adapt to change, and continue our work with renewed determination.

A New Norm to Embrace

For the past decade, we've witnessed a fast-paced change in terms of models and a deep challenge of the traditional patriarchal society. Some of these initiatives and new mentalities can seem overwhelming to some, illustrated by tensions around wokism, but behind these surges and more vocal cases lies a robust emerging trend. We are challenging certain mental constructions of predation in many areas of our lives, from the predation of nature and over-exploitation of resources to sustainability and regenerative practices and from the use of power over people for exclusive and selfish benefits to inclusion and distributive practices.

These changes constitute the foundations of new norms to come. It will probably take a few iterations and back-and-forths until each community can find its own balance, but the movement won't be stopped. Ecofeminism is an interesting concept that illustrates some of these models we will tend toward, and even if we were to adopt a fractional and diluted version of it, we would still be questioning management practices and how we interact with each other in different spaces. Among these principles lies the idea of antipredation, which asserts that one should not thrive at the expense of others or by depleting shared resources, known as the *commons*—such as public goods, natural ecosystems, or collective knowledge. Here again, vulnerability will play a big role by rebalancing the incentives and ethos in human groups and opening up new conversations aiming at the good of the whole group.

A Common Ground for Generations

As a teacher, I am at the forefront of this new generation, and I am both excited for my students to enter the job market and concerned about the older generations finding ways to work together positively. One common thread I've witnessed across generations is a poor acceptance of vulnerability, which will lead to conflicts and unsolvable situations. Vulnerability is the first and the best tool to make a step toward others, accepting that one generation is not better than the other.

Conclusion

Vulnerability is not a sign of weakness but a testament to our humanity. It is the courage to show up and be seen when we have no control over the outcome. Embracing vulnerability can lead to innovative ideas, more effective leadership, and increased resilience. Thus, in driving transformative change, vulnerability is not just a tool to be used but a mindset to be embraced. By doing so, we foster environments where creativity and empathy thrive, ultimately leading to more equitable and sustainable solutions to the challenges we face.

The Importance of Creating a New Narrative for Our Society

A significant barrier to progress and innovation often stems from the narrative we inhabit, also known as a *metanarrative*. This overarching story defines the playing field and sets the rules of the game in many ways.

A metanarrative, or "grand narrative," is an expansive, overarching story or theory that a culture, society, or group uses to explain and justify its beliefs, practices, and values. These narratives provide the framework through which individuals understand the world and their role within it. By offering a comprehensive explanation of historical events, social practices, and cultural phenomena, a metanarrative suggests that these elements are part of a larger, predetermined path or purpose.

Metanarratives profoundly shape the way people live and how societies function, influencing every aspect of our lives. Here's how they exert their power.

Metanarratives mold the collective worldview of a society, offering a lens through which people interpret their experiences. For example, the idea of progress in Western civilization is a metanarrative that posits history as a linear journey toward improvement and enlightenment. This belief permeates education systems and economic policies, driving the relentless pursuit of innovation and development.

These narratives often serve to legitimize existing social structures and power dynamics. During colonial times, for instance, the metanarrative of Western superiority was used to justify the domination and exploitation of other cultures. Similarly, the capitalist metanarrative promotes the belief that free markets and competition naturally lead to prosperity, shaping economic policies and societal values around profit and consumption.

On a personal level, metanarratives provide meaning and direction. Religious metanarratives, for example, often shape moral codes and influence daily decisions. In Christianity, the metanarrative of salvation and eternal life guides believers in their choices, actions, and understanding of their purpose in life.

Metanarratives also contribute to the formation of cultural identity by offering a shared history and collective purpose. Nationalism, for instance, is a metanarrative that unites people through a common national identity, often reinforced by stories of shared struggles, achievements, and values.

As societies evolve, challenges to dominant metanarratives can lead to significant social change. The feminist movement, for instance, challenged the metanarrative of patriarchy, advocating for gender equality and leading to profound shifts in societal roles, laws, and attitudes toward women. This revision maintains the original structure but refines the language for clarity and impact.

Transforming a metanarrative has the potential to create profound changes across all levels of society.

In many contemporary societies, the prevailing narrative glorifies the self-made individual who ascends to success through competition. This narrative often includes themes of domination over nature, where success is measured by the ability to extract and exploit natural resources, often leading to environmental degradation. It also involves domination over people, where competitive success frequently comes at the expense of others, resulting in significant disparities in wealth and power. Furthermore, this narrative tends to celebrate individual achievement while overlooking the collective efforts and systemic support that contribute to that success.

Although these modern metanarratives have driven remarkable progress in various areas, they may now need rethinking in light of our growing understanding of sustainability and social equity.

The Need for a New Narrative

If we remain anchored to a narrative where success is built on these principles, efforts to create a more equitable society will consistently face obstacles. What we need are new narratives that promote collective well-being, where individuals are valued not only for their personal achievements but for their contributions to the greater good. These narratives should celebrate sustainable practices and encourage collaboration over competition, underscoring the importance of working together to address common challenges.

Shifting a metanarrative may seem daunting, a task requiring years and the efforts of many people. However, there are countless ways we can each influence these narratives in our own lives. By doing so, we can contribute to shaping a more inclusive and equitable society, creating a progressive environment where people can thrive by supporting each other and the planet.

Redefining Success: A Personal Journey Toward a New Narrative

Challenging a metanarrative often begins with rethinking its outcome—success. By being intentional about how we define and measure success, we can influence not only our own motivations but also the aspirations of those around us. This, in turn, helps to reshape societal goals and the methods we use to achieve them.

During the pandemic, I found myself grappling with a deep sense of eco-anxiety and a strong desire to do more to protect the environment. This led me to make a series of deliberate changes in my life aimed at reducing my carbon footprint. I became a vegetarian, chose not to own a car (despite raising three young children in the United States), purchased only second-hand clothes and gear, and limited my travel—something that was relatively easy during the lockdowns of 2020. I even rented a plot in a local community garden where I grew my own food and cared for a beehive.

Two other changes were perhaps the most surprising. First, I decided to reduce my salary, not only because the economic situation was challenging for my company but also because I knew that our carbon footprint is closely tied to our income level. Second, my family and I moved to a smaller apartment, just two doors down from our previous home, but half the size. This transition wasn't easy at first. We had to part with many belongings and adjust to a more compact living space. Yet there were also significant benefits: our energy bills were halved, we felt better living with less, and the family became more closely bonded.

At the core of these decisions was the need to convince myself, my wife, and those around us that these were not sacrifices but rather desirable choices—that this was, in fact, a new definition of success. I had to change the narrative to one where success is about living with less, spending less on superficial things, enjoying what we have, and reducing our dependence on material possessions. Happiness doesn't come from owning more; a larger house or a faster car doesn't equate to a happier life. Often, it's quite the opposite.

This idea extends to the principles of inclusiveness and equality. By redefining success at our own scale and proposing new avenues for happiness, we can collectively shift the narrative.

Another action I took was to dedicate my Friday afternoons to volunteering with a local community organization called Soccer Without Borders. The organization uses soccer to help integrate underserved youth who have recently moved to the United States. Its mission is to create positive change for individuals, families, and communities through the power of soccer, providing these youth with the tools to overcome obstacles and achieve personal success.

Volunteering there was a humbling experience. Even though my role was as a coach, I often felt that I gained more from the kids than they did from me. Inspired by this experience, I began sharing it with those around me to raise awareness about the organization. I included this experience in my CV, LinkedIn profile, and other personal presentations, placing it alongside my roles as an angel investor and startup mentor. For me, this was a statement about what success should look like and how we should consider our impact. Supporting the personal development of a young refugee can be more impactful and fulfilling than investing thousands of dollars in a cool tech startup in Silicon Valley.

Success shouldn't be measured by elite status, personal wealth, or the busy schedules of hyperactive achievers. Instead, true success lies in participating in community organizations, nurturing a wide circle of loving friends and family, and embracing a pace of life that prioritizes quality human interactions and the generous use of our time. This approach not only fosters a more inclusive and equitable society but also leads to better health and overall well-being.

Promoting Visibility of Diversity in Influential Media

We often underestimate the profound influence that mass media has on our perceptions, behaviors, and beliefs. The images and narratives we see on television, in movies, and across social media platforms shape our understanding of what life is and should be. The characters, stories, and interactions we consume feed directly into the metanarratives that govern our society—subtly dictating who we aspire to be, how we act, and what we value.

These media representations mold our desires and attitudes, playing a significant role in shaping societal norms and expectations. The presence or absence of diverse voices and perspectives in these narratives can either reinforce existing biases or challenge and expand our collective worldview. For this reason, it's crucial to consciously seek out and promote diversity in the media we consume. By broadening our exposure to different perspectives, we can begin to shift the metanarratives that define our lives and the lives of future generations.

As parents and educators, this responsibility is even greater. The media we expose our children to has the power to shape their understanding of the world in fundamental ways. By ensuring that they encounter a wide range of voices and experiences, we help them develop a more inclusive and empathetic worldview.

At a broader societal level, we should encourage influencers, filmmakers, and the entire media industry to provide greater visibility to diverse people and ideas. This isn't just about representation; it's about challenging the dominant narratives that often marginalize or exclude certain groups. We've already seen positive strides in this area with the increasing visibility of LGBTQ+ characters in mainstream media, which has helped to normalize these identities and foster greater acceptance.

But influential media is not limited to large producers and TV channels; it can also include very immediate information channels like the class we are attending or the social media accounts we are following.

In 2020, as I was ramping up my Deplastify the Planet program—a unique open innovation initiative bringing together students from UC Berkeley with corporate partners to tackle plastic pollution—I had a personal encounter that underscored the importance of embracing diverse perspectives. The program was gaining traction, and I was actively promoting it to attract more participants. After sharing a post about some of our completed projects, I received overwhelmingly positive feedback, but one comment stood out. It was from someone who criticized the information I had shared, urging me to take it down.

As is often the case on social media, the comments quickly escalated, with friends and colleagues rushing to defend me and the program. But instead of dismissing this critic, I decided to dig deeper into his profile. I discovered that he was a scientist who spent much of his time online challenging initiatives aimed at reducing plastic use. He described himself as someone who helps people understand the realities of plastic.

Despite warnings from my colleagues, I invited him to be a guest lecturer in the program. The discussion was intense, and my students were visibly puzzled by his pro-plastic stance. When he left, the room fell silent. We spent the next hour unpacking his arguments, discussing what had just happened, and examining our own assumptions. Although he never returned to the class, the experience was transformative for everyone involved, myself included.

For once, my students were confronted with a radically different perspective, not as passive learners but as active participants in a debate that directly challenged their beliefs. They were treated not just as students but as economic agents by someone with a vested interest in influencing their opinions.

This experience reminded us all of the value of seeking out diverse viewpoints, particularly when learning about new and complex issues. It underscored the importance of questioning the metanarratives we take for granted, especially in the context of Inclusive Design. By diversifying our sources of inspiration and embracing a wider range of perspectives, we can develop a more balanced and inclusive narrative that better reflects the diversity of human experience.

Conclusion

Creating a new narrative for our society is crucial for fostering a more equitable and sustainable world. By shifting our stories from competition and predation to collaboration and collective well-being, we can redefine what it means to be successful. This new narrative will require not only a change in mindset but also the development of new metrics to measure and celebrate success. As we move toward these new narratives, we can create a society where everyone has the opportunity to thrive and success is shared and sustainable.

Notes

[1] Oxfam International, 2023. Richest 1% bag nearly twice as much wealth as the rest of the world put together over the past two years. [Online]. Available at: https://www.oxfam.org/en/press-releases/richest-1-bag-nearly-twice-much-wealth-rest-world-put-together-over-past-two-years [Accessed: 15 January 2025].

[2] Norton, M. I. & Ariely, D. Building a better America—one wealth quintile at a time. *Perspectives on Psychological Science*, 6(1), 9–12.

[3] Brown, B., 2018. *Dare to lead: brave work. Tough conversations. Whole hearts.* Random House.

Systempathy—or How to Empathize with People and the System They Live In

Inclusive Design is not just about addressing individual needs; it is about understanding the intricate systems in which people live, work, and interact. At the heart of this concept lies *systempathy*—a term I coined as a fusion of *system* and *empathy*, defining a deliberate empathy that considers not only individuals but the interconnected structures and norms that shape their experiences. This chapter delves into the transformative potential of combining empathy with systemic thinking, moving beyond isolated solutions to address the root causes of exclusion and inequality.

To create meaningful and lasting change, we must shift our focus from intention to impact. This requires more than good intentions; it calls for a nuanced understanding of how every design decision interacts with larger forces—cultural, historical, social, and economic. In the following sections, we'll explore how systemic thinking can guide us in crafting solutions that resonate deeply with individuals while driving change at a societal level. By adopting this holistic approach, designers, leaders, and citizens alike can contribute to a more inclusive world where systems evolve to meet the diverse needs of all.

From Intention to Impact: The Systemic Approach to Inclusive Design

Inclusive Design starts with noble intentions—a desire to create products, services, and experiences that are accessible, equitable, and impactful for all. However, translating these intentions into real-world change requires more than just a surface-level approach. It demands a deep dive into the complex, interconnected systems that shape our society, highlighting the importance of systemic thinking in design. Systemic thinking allows us to see beyond individual needs and behaviors, recognizing that each design decision interacts with larger forces, such as cultural norms, historical contexts, and socioeconomic structures. By embracing this broader perspective, we ensure that our solutions are not just isolated fixes but contributions to the greater goal of social equity.

In this chapter, we introduce the Inclusive Benefit Matrix, a practical tool that integrates both design principles and systemic thinking. It helps define the "why," "how," and "what" of any design initiative, offering a pathway to ensure that solutions are intentional, meaningful, and aligned with the broader ecosystems they inhabit. By leveraging this matrix, we can drive positive change not just for individuals but for communities and entire systems, transforming good intentions into lasting impact.

Systemic Thinking: The Foundation of Inclusive Design

At its core, Inclusive Design is about addressing the complex, interconnected challenges that individuals and communities face. These challenges rarely exist in isolation—they are often deeply embedded in the social, cultural, economic, and environmental systems that shape people's experiences. To create truly inclusive solutions, we must adopt a mindset that embraces this complexity—this is where systemic thinking becomes essential.

Systemic thinking is an approach that looks at problems holistically rather than in isolation. It acknowledges that individual actions, products, or services are part of larger systems, and any change to one part of the system can have a ripple effect on other parts. In the context of Inclusive Design, this means understanding that the products we design, the experiences we create, and the processes we develop do not exist in a vacuum. They influence and are influenced by a wide array of factors—social hierarchies, economic disparities, cultural norms, power dynamics, and more.

A simple everyday example illustrates this well: imagine an office kitchen where spilled coffee frequently appears on the counter. At first glance, the problem seems to be individual carelessness—people forgetting to wipe up after themselves. But taking a systemic perspective reveals deeper causes. Perhaps the coffee cups are too small, leading to frequent overflows. Maybe the machine is positioned awkwardly, making spills inevitable. Or perhaps there are no paper towels nearby, discouraging immediate cleanups. Instead of blaming individuals, a systemic approach would involve redesigning the setup—providing better-sized cups, repositioning the coffee machine, or making cleaning supplies more accessible. The same principle applies to Inclusive Design: what may seem like an issue of personal responsibility often stems from structural design flaws that, when addressed, create a more functional and accessible environment for everyone.

The theory of systemic thinking has its roots in multiple disciplines, but its modern development began in the early 20th century with the rise of fields such as systems theory, cybernetics, and ecology. One of its key pioneers was biologist Ludwig von Bertalanffy, who introduced General Systems Theory in the 1940s. This theory sought to explain how different parts of a system work together, influencing and being influenced by one another in ways that could not be fully understood by examining each part in isolation. Systemic thinking was further developed in areas such as organizational management, where it became a popular framework for addressing the complexity of human organizations, and in environmental science, where it helped shape our understanding of ecosystems and sustainability.

Today, systemic thinking is widely applied in fields such as urban planning, healthcare, business strategy, education, and technology. It is used to address "wicked problems"—issues so complex and multifaceted that they defy simple solutions, such as poverty, climate change, and inequality. By understanding the interdependencies within a system, systemic thinking allows us to anticipate unintended consequences and design solutions that foster resilience, adaptability, and sustainability.

Consider, for example, the experience of a user with disabilities navigating a public space. The challenges they face might not just be about physical accessibility but could also involve broader issues like societal attitudes, legal frameworks, and economic opportunities. If we focus only on designing a ramp without addressing the deeper, systemic issues—like social stigma, employment discrimination, or inadequate public services—we may improve access to a single building, but we are far from creating an inclusive society.

Systemic thinking encourages us to zoom out to see the bigger picture. It helps us identify the root causes of exclusion rather than merely addressing symptoms. For instance, when designing for underserved communities, it's not enough to simply provide a product or service; we must also consider how factors like trust, historical inequalities, or access to education play a role in the adoption and success of those solutions.

Moreover, systemic thinking pushes us to recognize the interdependencies within the systems we work in. A change in one area—whether through a product redesign or a policy shift—can lead to unintended consequences elsewhere. Inclusive designers must be mindful of how each decision affects the broader system.

This requires an ongoing cycle of research, observation, and iteration to ensure that our solutions do not inadvertently create new problems while solving existing ones.

Incorporating systemic thinking into the design process also forces us to confront the limitations of our own knowledge and perspective. No one person can fully understand the vast intricacies of the systems they are designing for, which is why collaboration and diverse perspectives are critical. Inclusive Design flourishes when we bring together individuals from different backgrounds and areas of expertise, enabling us to see the system from multiple angles and discover innovative, sustainable solutions.

By integrating systemic thinking into the heart of the design process, we can move beyond surface-level fixes and begin to create lasting, transformative change. It's about more than just building better products—it's about reshaping the systems that dictate how people experience the world, ensuring that no one is left behind in the process.

The Inclusive Benefit Matrix

An Inclusive Design process always begins with a clear intention. In most of the projects I've worked on, the intention was fairly straightforward, but the ultimate goal often remained vague and the path to achieve it even more elusive.

Inclusive Design and Equitable Design are still relatively new concepts, even though the principles of diversity, equity, and inclusion (DEI) are more familiar. This makes us all sensitive to the overarching objectives of social justice and mutual care. Initiating a conversation about Inclusive Design has always been relatively easy, regardless of who I'm speaking with or which industry they're in. Most people have genuine and well-meaning intentions when it comes to supporting a DEI-related cause. Whether it's because they support the LGBTQ+ movement, have a close friend or family member living with a disability, or belong to a minority group themselves, there's usually a personal connection to some form of social inequality that resonates with them.

The "why" behind these intentions is often the easiest part to address. The real challenge arises when we shift the conversation to "how."

Every week, I meet with a new person to discuss how Inclusive and Equitable Design can be applied to their specific context. These meetings are often with corporate professionals intrigued by the concept and driven by a desire to make a difference but unsure of how their role fits into the equation. Initially, many of them feel that although they understand the concept, it doesn't seem relevant to their specific industry or job. But by the end of the conversation, they usually come to realize that, in fact, they can create an impact, even at their own scale.

In 2022, I met with a long-time client from a personal care brand. We had previously collaborated on several sustainability-focused projects, and this time

I wanted to introduce the concept of Equitable Design. Their initial response was, "That sounds great, but I don't think it applies to us. We're just selling shampoo and detergent."

The "how" wasn't immediately obvious to them. But that's when they realized they needed help to explore this further, and we embarked on a fantastic project together, discovering how something as everyday as shampoo could become a vector for inclusion and equality. I'll share more details of that project in a next chapter.

I've had similar experiences with leaders from industries as diverse as alcoholic beverages, notebook manufacturing, developer software, entertainment, insurance, blockchain, and champagne. In each case, there was an initial skepticism about how Inclusive Design could apply to their field, followed by a realization of its potential once we dove deeper into the possibilities.

From these interactions, I developed a simple tool to help clarify the goals behind the intention: the Benefit Matrix (see Figure 3.1).

The Benefit Matrix is composed of two axes. One axis moves from universal design to personalized design, and the other ranges from reducing disadvantage

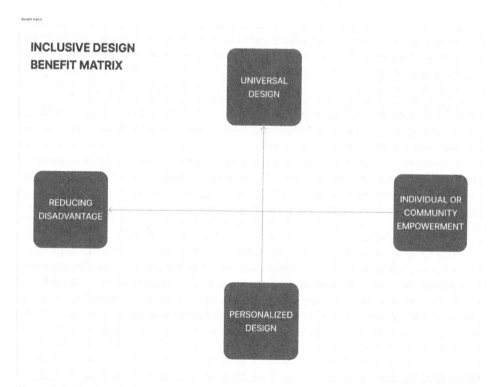

Figure 3.1: The Benefit Matrix

to individual and community empowerment. This creates four quadrants, each representing a potential goal or approach.

For instance, when working on a consulting project with students and a major social media platform, we developed a personalized design to empower local Black business owners to advertise on a new media format. The interface was carefully tailored to this specific group, incorporating cultural references that resonated with their communities. As a result, these businesses were more inclined to use the platform to promote their services, achieving higher audience engagement. This increased visibility and positive reputation not only drove more business but also helped strengthen their communities through enhanced connections and opportunities.

On the opposite side of the matrix, we collaborated with a top IT company to revisit its authentication process, making it more inclusive for users with visual impairments who struggled with typing passwords or dictating them aloud in public. The goal was to reduce their disadvantage while offering a universal experience that could work for all users.

This matrix has proven to be an effective tool, guiding companies and change-makers in determining the goal that best aligns with their "why" and helping them translate their intentions into meaningful action.

At the same time, it is important to recognize that the path toward achieving inclusivity is rarely straightforward. The decisions that shape the direction within the matrix are often context-dependent and require balancing the diverse needs and interests of stakeholders. Aligning the expectations of immediate users, excluded users, shareholders, product teams, and surrounding institutions is complex and often involves trade-offs. Although the initial intention is almost always noble, the execution may demand compromises, adjustments, and acceptance of imperfect solutions.

My own experience includes many projects where we encountered such challenges—moments where we had to reconcile conflicting priorities, adapt to unforeseen constraints, or acknowledge that our efforts might not achieve the level of impact we had hoped for. The following four sections illustrate projects that unfolded smoothly and delivered on their promise. However, behind these successes lie equally important experiences where difficulties emerged, pushing us to learn and evolve. These challenges are an integral part of the Inclusive Design journey and serve as opportunities for critical growth and insight.

Delving deeper into the concepts of this matrix, we begin to grasp the systemic approach that underpins Inclusive Design and establish the foundation for the diagnostic work that must be undertaken at the outset of any project.

Universal Design

As discussed in Chapter 1, "The Fundamentals of Inclusive Design," the concept of universal design is far from new, and in many cases, it is seen as a mandatory requirement by many of the people I work with. The idea of creating a product or service that accommodates so-called extreme users while compromising the experience for mainstream users may seem contradictory to the goals of Inclusive Design. For businesses, there's often a legitimate concern that altering their product, service, or process to be more inclusive might alienate current customers.

The essence of universal design is to create an experience that has no barriers to access, and in many instances, this approach works exceptionally well.

Take, for example, a project I worked on with a leading IT company to redesign its authentication process. Initially, the service was quite fluid for the average user: when logging in to their professional accounts on a computer or phone, users were prompted to authenticate via a standalone app developed by the company. This app generated a six-digit code that users entered on the original interface.

However, we quickly identified a significant issue for users with visual impairments. Reading and transcribing a code from one screen to another was not an option for them. Instead, they had to rely on a voice assistant to read the code aloud and then repeat the code out loud to enter it—a process that's far from secure, especially in public spaces like libraries or offices.

The challenge was to design a new authentication process that would maintain security for visually impaired users while ensuring a seamless experience for sighted users—achieving a true universal design solution.

Our team developed a secure system that used geolocation to confirm the identity of the person attempting to log in, eliminating the need for a passcode altogether. The result? A process that not only maintained security for all users but also shaved up to 28 seconds off the authentication time. This solution was beautifully simple and efficient for everyone—a perfect example of universal design in action. I'll elaborate further on this solution and its business impact in a later chapter. Similarly, many examples provided in the following chapters will be revisited to illustrate key concepts along the way.

Personalized Design

Unfortunately, universal design isn't always feasible. Consider a simple scenario: designing a chair that accommodates both short and tall people. The challenge presents itself as a conflict—what works well for one group may not work for the other. This dilemma arises when designing products or services that incorporate significant cultural elements. It's impossible to create a single solution that is truly inclusive of all cultures while being universally the same for everyone.

In such cases, we recommend creating a range of products, each tailored to the specific needs and preferences of different target audiences—a strategy known as personalized design.

One example comes from a program we conducted with a large fintech (financial technology) company. Although its entire design and product team was based in San Francisco, its clients were scattered across the globe, representing a diverse range of cultures. When dealing with finance and money, trust and symbolism play a crucial role in the user experience. After receiving some negative feedback about the platform's user experience in certain Asian countries, we embarked on a journey to create a more locally authentic experience that fostered trust among the company's users.

We delved into the unique cultural contexts of its target markets, analyzing local beliefs, customs, and the significance of colors and symbols, as well as the importance of language and formulation.

For instance, we discovered that in Chinese culture, the color red symbolizes good luck and prosperity. Red is traditionally used during special occasions such as the Lunar New Year or weddings, where money is given in red envelopes as a symbol of fortune and success. However, in many Western cultures, a red background on a screen conveys danger or failure. These contrasting interpretations required careful consideration in the design.

In Indian tradition, Thursday is considered an auspicious day for financial transactions and starting new ventures. This belief is tied to the planet Jupiter, which is associated with wealth and fortune. With this insight, we adjusted the company's marketing strategy, sending email campaigns for financial services on Thursdays to resonate with local users.

Armed with these insights, we developed different versions of the fintech platform, each tailored to the cultural context of its target audience. This highly personalized approach allowed the company to build trust and engage users in ways that felt relevant and authentic to their local values and traditions.

Reducing Disadvantage

In many cases, the primary goal of Inclusive Design is to ease access to a service, with the intention of leveling the playing field and reducing disadvantages for certain audiences.

One notable project involved working with a leading entertainment company. Its user interface was highly effective and visually appealing, designed to generate excitement for users as they browsed through movie jackets and trailers on their streaming platform. However, the company wanted to focus on making the experience more accessible to users with visual impairments. For these users, the interface—although thrilling for the general audience—was overwhelming and difficult to navigate. Audio descriptions, which read out every item on the screen, often resulted in sensory overload, making it nearly impossible for visually impaired users to enjoy a seamless experience.

We completely reimagined the browsing experience, streamlining the decision-making process and optimizing the audio prompts to guide users more efficiently. Instead of bombarding them with unnecessary information, the new design allowed users to quickly and intuitively get to the movie they truly wanted to watch.

The goal of this redesign was not just to improve the aesthetics or functionality of the platform but also to reduce the disadvantage that visually impaired users faced. By simplifying the interface and providing clear, well-structured audio navigation, we made it possible for them to browse and select movies in a reasonable time, thus enhancing their overall user experience.

Empowering Communities

In some cases, Inclusive and Equitable Design work goes beyond reducing disadvantages and instead aims to uplift and empower specific communities. The goal here is not merely to bridge gaps between different groups but to provide communities in need with tools and resources that can significantly enhance their influence and opportunities.

One such example involved supporting a major social network in its efforts to help local Black business owners advertise on a new media format. This initiative came about just months after the George Floyd incident, as the company sought meaningful ways to support and empower Black communities across the country. Although it was already engaged in efforts like increasing diverse hiring and making donations to NGOs and charities, we proposed a more sustainable and impactful approach: turning the company's existing products and services into growth drivers for African American communities.

Rather than creating new, costly programs, the idea was to leverage the company's existing tools—particularly its advertising platform—to empower Black business owners. By redesigning the experience of this new media format specifically for Black business owners, we aimed to make it easier for them to use and benefit from the platform. The goal was to help these businesses generate greater visibility, attract more customers, and ultimately drive more revenue.

The beauty of this approach was that it created long-term, sustainable value for the community at no additional cost to the company. By empowering Black business owners to take full advantage of the platform, we were able to enhance their ability to create jobs, strengthen their reputation, and build resilience within their communities.

This project illustrates the power of designing not just to serve but to empower, creating tools that allow communities to uplift themselves and thrive.

Connecting the Dots: The Systemic Impact of Inclusive Design

When examining the matrix, it becomes clear that Inclusive Design isn't just about enhancing individual user experiences or increasing revenue. It's about breathing life into an intention that operates within a broader system. When we aim to empower communities, we are challenging the status quo of a particular system. Our goal is to create more opportunities for people within a specific ecosystem, and to do so effectively, we must first understand the intricacies of that ecosystem and identify the leverage points that can help shift the balance of power.

Take, for example, the case of empowering Black business owners through a social media platform. The social media tool is merely an enabler within a much larger system. By giving these businesses prime access to a broader social media audience, we're increasing their visibility, which in turn nudges more customers toward their products or services. This leads to higher revenue, which translates into job creation and new opportunities within that community. The impact ripples outward, but only if the system as a whole is taken into account.

Similarly, when striving for universal design, we must consider not only our target users but also the broader audience we are not specifically designing for. This requires a systemic perspective to ensure that as we serve one group, we do not inadvertently exclude others. Universal design, by definition, should be accessible to as many people as possible, which means paying attention to the entire ecosystem rather than focusing solely on a small group of individuals.

Moreover, finding the most effective solution often comes from looking at the entire system surrounding a product rather than just the product itself. For instance, when redesigning the authentication service, many designers might have concentrated solely on refining the existing interface. However, by considering the broader system—such as the multiple devices and technologies that users interact with—we were able to develop a more holistic solution. This approach underscores the importance of viewing Inclusive Design through a systemic lens, ensuring that every component works together to deliver an experience that benefits everyone involved.

Defining the Problem: The Foundation of Successful Design

Now that we have clarified the "why" and the "how" of the project—the "why" is the initial intention behind the Inclusive Design initiative, and the "how" is generally determined by the quadrant of the Benefit Matrix (refer to Figure 3.1)—it's time to focus on the "what" and delve deeper into our approach. The most critical aspect of any design project is correctly identifying the problem we are trying to solve. I often say that if you create a solution without addressing a real problem, that solution will quickly become your next problem! Too often, innovators and changemakers get excited about a solution—either because they've advanced a particular technology or they are trying to replicate success from another industry (how many times have we heard about "the Uber of X industry"?). Yet they rarely stop to ask the fundamental question: "What problem am I actually solving?"

This relentless focus on identifying the right problem transforms Inclusive Designers into experts within their domain, driving a deep understanding of the environment in which they operate. Solving the right problem accounts for 90% of the innovation effort, as this phase is where the crucial expertise and insights are developed, laying the foundation for a truly effective solution. Conversely, a poor assessment of the problem likely accounts for more than 90% of innovation failures. Without a clear and accurate problem definition, even the most sophisticated solutions are doomed to fall short.

The Role of Empathy in Inclusive Design

Design Thinking is an essential part of the innovation process, often referred to as the "thinking" phase within the broader scope of design. This methodology emphasizes a human-centered approach, integrating the needs of people with the possibilities of technology and the requirements for business success. Central to Design Thinking is the concept of empathy—a vital element that fosters a deep connection with the end user. Empathy in design goes beyond simply acknowledging users' needs; it involves immersing oneself in their experiences, emotions, and challenges. This deep understanding allows designers to create solutions that truly resonate with users, addressing not just functional requirements but also the subtler aspects of user experience.

Understanding Empathy in Design

Empathy is the ability to put oneself in another's shoes, to see the world through their eyes, and to feel their emotions. In the context of design, empathy involves immersing oneself in the user's environment, observing their interactions and listening to their stories to gain a profound understanding of their experiences. This empathetic approach enables designers to identify the real problems users face and to create solutions that genuinely address their needs.

Jeremy Rifkin, in his book *The Empathic Civilization* (Polity Press, 2010), argues that empathy is a fundamental human trait that has the potential to transform society. He writes, "Empathy is the invisible hand. It's what bonds societies together and allows us to advance. When empathy flourishes, society flourishes." Rifkin's perspective highlights the critical role empathy plays not only in societal development but also in innovation and design.

In the Empathize stage of a design process, designers immerse themselves in the user's world. This involves techniques such as shadowing, where designers follow users throughout their daily activities to observe their interactions and behaviors firsthand. User interviews also play a crucial role, allowing designers to ask open-ended questions that reveal the emotional and experiential aspects of users' interactions with a product or service. Empathy mapping is another powerful tool used in this stage, helping designers visualize users' experiences, needs, and emotions, providing a holistic view of their lives.

Empathy enables designers to move beyond assumptions and biases, fostering a deeper connection with users. This connection is crucial for several reasons:

Revealing unarticulated needs   Users often cannot articulate their needs or problems clearly. Empathy allows designers to observe and infer these needs through users' behavior and emotions. For example, IDEO, a global design company, used empathy-driven research to redesign a children's hospital experience. By observing the hospital journey from a child's perspective, the company identified pain points and created solutions that significantly improved the experience for young patients and their families.

Fostering innovation   Empathy-driven insights can lead to breakthrough innovations. When designers understand users' struggles and aspirations, they can create solutions that users didn't even know they needed. The development of the Swiffer, for instance, came from Procter & Gamble's designers immersing themselves in people's cleaning routines. They discovered that traditional mops were cumbersome and ineffective, leading to the creation of a more user-friendly cleaning tool.

Building emotional connections   Products and services designed with empathy resonate more deeply with users, creating emotional connections that foster loyalty and satisfaction. Apple exemplifies this through its relentless focus on user experience and intuitive design. The company invests heavily in understanding how users interact with technology, resulting in products that feel effortless to use. By paying close attention to design details—such as the haptic feedback on the iPhone or the seamless integration of hardware and software—Apple creates a sense of delight and trust that keeps customers returning.

Not all players use the same methodologies, but the ability to evoke positive and intense emotions toward something as seemingly simple as a piece of aluminum and glass is a testament to a deep understanding of user sensitivities and connections. This demonstrates the transformative power of empathy in design—turning functional objects into meaningful experiences that resonate on a personal level.

Avoiding confirmation bias   Empathy in design is often cited as a cornerstone of the creative process, but truly practicing it is far from straightforward. It requires a conscious effort to step outside one's own perspective and engage deeply with the experiences and emotions of others. This process can be challenging because designers may unintentionally fall into the trap of confirmation bias, where the empathy phase becomes more about validating preexisting assumptions than genuinely exploring the user's world.

For example, during user interviews, questions can be subtly or overtly leading, steering the conversation toward answers that align with what the designer expects or hopes to hear. Instead of uncovering fresh insights, the process may reinforce existing beliefs, limiting the potential for innovation. It takes deliberate effort and self-awareness to craft neutral, open-ended questions that allow users to express their thoughts and feelings freely, without influence.

Consider a scenario where a team is designing a new healthcare app. If the interview questions are framed around assumptions that users are primarily concerned with ease of use, the responses will likely confirm this bias. However, if the designers approach the interviews with an open mind, asking broad questions about users' overall experiences with healthcare, they may uncover deeper concerns, such as anxiety about medical data security or frustrations with the lack of personalized care. These insights could lead to more meaningful and impactful design solutions.

In essence, empathy in design is not just about asking the right questions but also about creating an environment where true exploration can occur. This means being vigilant against biases, remaining open to unexpected insights, and genuinely listening to users' stories and experiences. It's a skill that requires continuous practice and reflection, but when done right, it can lead to innovations that are not just functional but also deeply resonant with users' needs and desires.

The Broader Impact of Empathy

Empathy in design extends beyond product development. It influences organizational culture and societal impact. Jeremy Rifkin highlights the potential of empathy to address global challenges, stating, "We are beginning to see the outlines of a new economic system that will allow us to share the planet's resources more efficiently and sustainably, and to move from individualistic to collaborative relationships." Empathy-driven design can contribute to this shift by creating solutions that consider the well-being of all stakeholders, promoting sustainability, inclusivity, and social equity.

Essential Tools for Assessing Needs and Challenges

To undertake an Equitable Design project, one often draws on methodologies from anthropology and ethnography, disciplines that provide deep insights into human behavior, culture, and social interactions. Anthropology, the study of humans in their cultural and social contexts, and ethnography, a research method focused on the systematic study of people and cultures through direct observation and participation, are invaluable tools in understanding the diverse needs of different user groups.

Anthropology and Ethnography in Design

Anthropology is a broad field that examines various aspects of human life, including cultural practices, social structures, language, and belief systems. By understanding these elements, designers can gain a more comprehensive view of the user's environment, motivations, and constraints. In the context of Equitable Design, anthropology helps to uncover the underlying cultural factors that influence how people interact with products, services, and systems.

Ethnography, a key method within anthropology, involves immersing oneself in the daily lives of people to observe and record their behaviors, interactions, and experiences. This method is particularly useful in design because it allows researchers to gather rich, contextual insights that go beyond surface-level observations. Ethnography often involves participant observation, where the researcher actively engages with the community they are studying, and in-depth interviews, where individuals share their experiences and perspectives in their own words.

When applied to design, these methodologies enable a deep understanding of users from diverse backgrounds. For example, in designing a healthcare service for a multicultural urban population, an ethnographic study might reveal significant differences in how various cultural groups perceive healthcare, seek medical help, and interact with healthcare providers. These insights could then inform the design of more culturally sensitive healthcare services that cater to the needs of all users rather than a one-size-fits-all approach.

In product design, anthropology and ethnography can help designers to move beyond assumptions and biases. For instance, while developing a new kitchen appliance, a designer might conduct ethnographic research in different households to observe how people cook, what tools they prefer, and the challenges they face. This research could uncover that certain cultural groups have unique cooking practices that are not well-supported by existing products, leading to the creation of a new appliance that better meets the needs of those users.

By integrating anthropology and ethnography into the design process, designers can ensure that the products, services, or processes they create are more inclusive and equitable. These methodologies help to identify the diverse needs of users, challenge assumptions, and develop solutions that are grounded in real-world experiences and cultural contexts. This approach not only leads to better design outcomes but also contributes to the broader goal of social equity by ensuring that all users, regardless of their background, are considered and valued in the design process.

All the Inclusive and Equitable Design projects I've worked on start with this effort to better understand user needs. We either begin with an existing product, service, or process or are set to create a new one. In any case, we know that some people will be affected by our design. Whether they will be direct users or indirect stakeholders, we want to gain a better understanding of who they are, what their lives and experiences around our focus area are, and how we could make them better. With Inclusive and Equitable Design, the goal is to have a product or service that would perform better but also to contribute to a more inclusive and equitable society. The range of our research must go beyond the boundaries of the experience we're working on and look at all externalities and contributions to exclusion and inequalities for any audience.

The stakes of this phase are pretty high because the needs and the context we will uncover will determine our design choice and, by extension, the impact of our solution, regardless of the next phases.

In this section, we will explore a few unique tools and approaches we are using to reveal valuable insights.

Be an Investigator, Not a Judge

I've often encountered situations where students present their research with absolute certainty, fully convinced of their findings. A common scenario involves groups confidently asserting that consumers would willingly pay a premium for products that are more expensive but fairer for the workers involved or for those supporting charitable causes. When they present this to me, brimming with assurance, I often ask, "What makes you think this?" Their response is typically, "We asked people—out of five interviews, four said they would do so." I follow up with, "And why do you trust what they say?" To which they reply, "Because they are our friends!" Then I pose the critical question, "So, when was the last time any of them actually bought a fair trade product or supported a product contributing to social initiatives?" This question is usually met with silence.

The reality is that although people often express good intentions, their actions don't always align with what they say. This gap between what people claim and what they actually do is crucial for designers to investigate, not judge. It's not that individuals are intentionally dishonest; rather, there's often a significant difference between theoretical intentions and practical actions. For designers, these gaps are rich opportunities, highlighting unmet needs and areas where existing products or experiences fail to meet consumer desires and aspirations. Understanding these discrepancies allows us to create more meaningful and impactful designs.

The empathy map in Design Thinking (see Figure 3.2) is a powerful tool designed to help teams deeply understand the user by mapping out not only what the user says and does but also what they see and hear. By using this grid, designers can gain a holistic view of the user's experience, enabling them to identify cognitive dissonance or unmet needs that may otherwise go unnoticed.

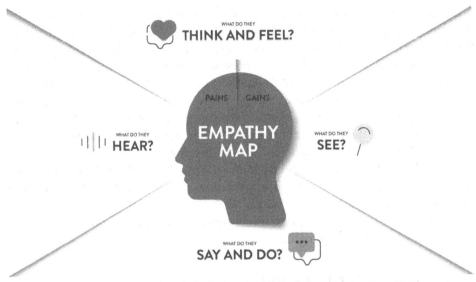

Figure 3.2: The empathy map, by whale design

This approach allows for a more nuanced understanding of the user's context, facilitating more effective solutions that address the underlying challenges and aspirations of the people they are designing for.

At the beginning of my career, I was supporting a startup that was developing an innovative headset equipped with sensors to track stress levels and offer tailored sound journeys to help users relax. As part of the design process, we needed to settle on the best color for the headset. The aim was to create something tech-forward yet stylish, appealing enough for people to wear comfortably in various environments, similar to Apple's sleek design approach.

The design team came up with several color options: black, gray, and white to cater to minimalist preferences and brighter shades like blue, green, and yellow for those who might want something more eye-catching. We gathered a small group of potential users for a focus group to get their thoughts on the color options. The session included 3D renderings of the headset, and we asked simple questions like, "Which color would you prefer for the headset?" After gathering feedback, the results were overwhelmingly clear—most participants preferred gray and black.

To thank the participants, we handed out iPod Minis (see Figure 3.3) at the end of the session. The iPods were neatly arranged on a table, available in five different colors, and everyone was invited to pick one. I watched with interest as each participant made their selection. To my surprise, the vibrant-colored iPods—reds, blues, and greens—were snatched up immediately, while the gray and black models remained untouched.

This was a perfect illustration of the disconnect between what people say and what they actually do. When asked in a theoretical setting about a generic

Figure 3.3: Apple campaign for iPod Nano-chromatic, displaying the range of bright colors: silver, black, purple, light blue, green, yellow, orange, red, and pink

product, the participants gravitated toward neutral, "safe" choices: they seemed to choose based on what they assumed the "silent majority" would prefer. However, when given the chance to pick something for personal use, their behavior reflected a completely different set of preferences. They opted for vibrant, bold colors when it came to something they'd carry every day—revealing the importance of context and emotional connection in design choices.

One of the most effective ways I've found to uncover the gaps between what people say and what they do is to diversify information sources and create highly contextual observation experiences. This approach allows me to compare different data points, helping to expose behaviors that wouldn't typically emerge in traditional interviews or surveys. As seen in the example of the headset focus group, where participants overwhelmingly preferred neutral colors during a questionnaire but chose vibrant colors when given the chance to select an iPod, context can lead to radically different behaviors.

Another instance occurred during a project focused on food products. In interviews, parents confidently claimed they regularly cooked healthy, vegetable-rich meals for their families. To validate this, I introduced an alternative data collection method: a daily request for dinner-time photos from these parents. The results told a different story. The majority of the meals captured were quick, frozen, or microwavable dishes, far from the nutritious meals described in the interviews. Here again, the participants were not lying but rather expressing their aspirations. However, in the rush of daily life with hungry children, the reality was that convenience often won out over healthy cooking.

This might seem contrary to the concept of empathy, as it may appear that the designer is not strictly listening to the user's expressed desires. However, it underscores that designers must transcend surface-level statements and focus on uncovering the users' deeper, unarticulated needs. True empathy involves not only hearing what users say but also understanding the realities and constraints they face. In this case, the designer's role is to align with the broader context of users' lives—acknowledging their aspirations while addressing the practical challenges that influence their behaviors.

In Inclusive and Equitable Design, we aim to create products and services that address deep-rooted issues related to accessibility and social equality. Although many of the traditional tools from mainstream design, such as observation and customer discovery, remain valuable in this space, the challenges we seek to solve go beyond surface-level convenience, like booking a taxi or ordering groceries online. Instead, we are tackling issues of exclusion, inequality, and systemic unfairness—problems that are inherently more difficult for users to articulate and for designers to fully grasp through simple observation or empathy exercises. These issues are often more nuanced and personal, requiring designers to go beyond merely stepping into the user's shoes. This is why it is critical to diversify sources of information and conduct deeper, more multifaceted research to uncover these hidden struggles and create truly inclusive solutions.

In 2024, I worked with a world-leading beverage company to create more inclusive celebration experiences for individuals with neurodivergence. Although conducting interviews was a key part of the research, my team and I took an additional step: we immersed ourselves in bars and social gatherings, observing interactions and behaviors in real-world settings. This contextual research revealed deeper insights than interviews alone could have uncovered.

For instance, we encountered two students who frequented a bar in the early afternoon not because they enjoyed the partying atmosphere of the bar but to collect loyalty points. They admitted they didn't resonate with the bustling evening crowd but were drawn in by the bar's gamified loyalty program. Interestingly, they shared that their lack of interest in nightlife sometimes led to feelings of illegitimacy, even embarrassment, as it went against the social norms of their peer group. By visiting the bar during off-peak hours, they found a way to experience the bar and foster a sense of belonging without the anxiety of crowds, subtly re-creating a connection to the broader student social circle.

We also met an individual with autism who grew visibly anxious before entering a bar, as they felt comfortable sitting at only one specific table—a quieter, more hidden space. These nuanced behaviors shed light on how physical space design can impact inclusion. Without direct observation, insights like these would likely remain undiscovered, as they go beyond what could be captured in traditional interviews or imagined by simply attempting to "step into the user's shoes."

This experience highlighted the importance of creating environments that cater to a broader range of needs and behaviors, ensuring that everyone can feel at ease in social spaces.

Whether it's discovering the truth behind people's purchasing decisions, recognizing the gap between their ideals and their realities, or understanding how individuals adapt to environments that aren't well-suited for them, these discrepancies offer invaluable insights into Inclusive and Equitable Design.

Inclusive Design and the Power of Status

Inclusion and social justice require us to examine more than just visible barriers; they demand a focus on subtler, often silent challenges such as those related to social status. These status-related challenges can be powerful, shaping how people perceive themselves and others in various contexts, particularly when interacting with products and services.

When people purchase or use a product, they are often sensitive to the message that the product conveys about them. A classic example can be seen with luxury cars like a brand-new Porsche. Driving such a vehicle sends a clear signal to those around you: it suggests wealth, sophistication, and perhaps a love for speed. The car becomes more than just transportation—it's a symbol of status, shaping the way others perceive you. Especially considering that in most cities around the world, speed limits make driving a race car model largely impractical!

Similarly, everyday choices like clothing also speak volumes. Someone who wears sportswear in most settings may project an image of being active, practical, and focused on comfort and fitness. These choices are not just about personal preference but about identity and social signaling—how individuals want to be perceived by those around them.

Products and services serve as extensions of a person's identity, communicating traits like wealth, style, or priorities to others. This plays a significant role in shaping social status. In the context of inclusion, this dynamic becomes particularly important because it highlights how certain products or services can either foster belonging or create division. For example, access to luxury goods can affirm one's place in an elite group, whereas the lack of access to similar goods might reinforce feelings of exclusion or inferiority.

Inclusion efforts must, therefore, account for these status dynamics. It's not just about making products available to everyone but also about ensuring that the experience of using those products doesn't inadvertently exclude or marginalize people based on their social standing. By addressing the status barriers built into the consumption of goods and services, we can create environments where people feel a stronger sense of belonging regardless of their economic or social background.

This understanding of status and its influence on even the most basic consumer decisions begins at a young age. I once collaborated with a major notebook company, trying to better understand how people made their purchasing decisions. To do so, I interviewed consumers from diverse backgrounds and accompanied them to stores to observe their choices firsthand.

One day, I joined Paul and Daniele, parents of two young children, on their back-to-school shopping trip. Paul and Daniele come from humble beginnings, neither having attended high school or college, yet they are well-respected in their community. Both work hard to ensure their daughters have the best possible opportunities, even though they believe their lack of higher education has limited their income and access to certain resources. They are typically very cautious about spending, avoiding unnecessary expenses.

As we walked through the busy aisles of the shopping center, I noticed something surprising: Paul and Daniele were consistently drawn to the more expensive products. At first, I thought this was a clear example of marketing at work—bright packaging and well-known brands influencing their decisions. But when I asked Paul about their preferences, he offered a more profound explanation: "This notebook looks better, more professional. I think it'll help my daughter learn better. Because I can't always help her with homework, the least I can do is give her the same tools as the best students."

Daniele chimed in, saying, "I know teachers pay more attention to privileged kids, and I don't want them to think we're poor or don't care enough. I want them to give our daughters the same chances."

This moment was a revelation. It showed how, for these parents, the appearance of something as simple as a notebook carried significant social weight. To them, it wasn't just about quality—it was about signaling status, ensuring their children weren't overlooked, and giving them the best possible start in life. Whether or not this belief was accurate, it underscored the deep social role that even the most basic objects can play in shaping perceptions and opportunities.

The issue of social status is a critical consideration when exploring innovation through Inclusive Design. It prompts us to ask: can we create solutions that don't merely reinforce existing hierarchies of social status within our field? And conversely, can we design products or services that empower individuals who might otherwise be deprived of social recognition? These are essential questions, as the potential of Inclusive Design lies not just in accessibility but in its ability to elevate those who are marginalized by offering them visibility, dignity, and participation through thoughtful, Equitable Design. By addressing these motivations, we can open up new avenues for creating products that resonate beyond function, positively influencing societal perceptions of value and status.

An interesting example of a brand that helps people from lower-income backgrounds project a higher social status is Telfar (see Figure 3.4), the fashion brand behind the now-iconic Telfar shopping bags. Often referred to as the "Bushwick Birkin," the Telfar bag has gained immense popularity not only for its stylish and sleek design but also for its affordable price point, which is a fraction of what high-end luxury bags typically cost.

Telfar's approach to fashion is both inclusive and empowering. The brand has democratized the idea of luxury by offering high-quality, trendy bags that are accessible to a wider audience, allowing people from various socioeconomic backgrounds to own and flaunt a product that signals style, sophistication, and a certain cultural status. Unlike traditional luxury brands that maintain exclusivity through high prices, Telfar has built its brand around inclusivity, with the tagline "Not for you, for everyone," further emphasizing that its products are made for all.

What makes Telfar's bags so influential is not only their design but also the cultural narrative surrounding them. Celebrities, influencers, and fashion enthusiasts alike carry Telfar bags, which has elevated the brand to a status symbol, yet one that is reachable for many. For individuals from lower-income backgrounds, purchasing a Telfar bag allows them to participate in a high-fashion narrative without the exorbitant cost, effectively using the product to project a higher social status.

By making its products accessible while maintaining cultural relevance, Telfar shows how Inclusive Design can challenge traditional luxury models and offer new pathways for individuals to assert their identity and status in society.

Figure 3.4: Telfar advertisement

The Weight of the Past in Inclusive and Equitable Design

Another dimension that is often overlooked in mainstream design but is absolutely key in Inclusive and Equitable Design is the weight of the past for certain users. Although traditional design methods tend to focus on the present-day needs of users, Equitable Design calls for a broader understanding of the psychological and spiritual factors that may shape an individual's experiences, perceptions, and behaviors. One such factor is the notion that individuals carry with them the legacy of seven generations, a concept rooted in both spiritual traditions and modern psychology.

The saying that we come to life carrying the legacy of seven generations suggests that trauma, experiences, and unresolved issues from our ancestors can shape our emotional responses and behaviors today. In various Indigenous and spiritual traditions, this belief holds that the effects of past events, such as trauma, fear, and conflict, are passed down through generations, impacting the well-being of descendants. This spiritual element implies a connection that transcends time, where healing the individual involves addressing not only personal experiences but also the unresolved traumas of those who came before.

From a psychological perspective, this concept has gained attention through research in the field of epigenetics and generational trauma. Studies have shown that trauma can have lasting effects on genetic expression, which may be passed down to subsequent generations. This phenomenon has been observed in descendants of Holocaust survivors, children of war victims, and communities that have endured systemic oppression. The inherited trauma can manifest in unexpected reactions, emotional responses, or physical symptoms, often without a clear present-day cause. Psychologically, individuals may carry the weight of these inherited experiences without conscious awareness, influencing their interactions with products, services, and environments.

The concept of Post Traumatic Slave Syndrome (PTSS) offers a profound lens through which to explore the generational legacy of trauma, particularly in the context of communities that have endured systemic oppression. Coined by Dr. Joy DeGruy, PTSS refers to the multigenerational trauma experienced by African Americans as a result of slavery, institutionalized racism, and oppression. Dr. DeGruy's work highlights the ways in which the psychological wounds from slavery are passed down through generations, influencing behaviors, emotions, and societal conditions long after the original trauma occurred.

PTSS suggests that certain survival mechanisms, such as hypervigilance and internalized oppression, that were developed during slavery and subsequent systemic oppression continue to affect descendants even today. This trauma manifests not only in psychological patterns but also in the social, emotional, and physical health of individuals and communities. Dr. DeGruy notes that these patterns are compounded by ongoing racism and discrimination, making it difficult for many African Americans to escape the cycle of trauma without deliberate healing efforts.

In the context of Inclusive and Equitable Design, this highlights the importance of acknowledging historical trauma as a key factor in understanding certain user behaviors and needs. The weight of this legacy can affect how people interact with products, services, and systems, and designers must be attuned to these deep-seated emotional and psychological experiences to create truly inclusive solutions.

For instance, when designing services for communities that have experienced historical trauma, it's crucial to consider that their reactions, hesitations, or needs may be influenced by inherited fears, mistrust, or other emotional responses rooted in past experiences. This also means providing spaces for healing, empowerment, and recognition of these histories within the design process rather than dismissing them as irrelevant.

Dr. DeGruy's research offers valuable insights into why certain communities may face unique challenges or exhibit specific behaviors that seem disproportionate to the present-day situation. It emphasizes the need for empathy and a deeper understanding of the historical and cultural contexts that shape people's lives, behaviors, and needs.

Equitable Design goes beyond addressing surface-level needs; it requires recognizing that users may be influenced by unseen forces, such as ancestral trauma or inherited emotional patterns. Designers must be mindful of the fact that, for some, the user experience is not just about convenience or functionality but is intertwined with complex, intergenerational emotions. By incorporating this understanding into the design process, we can create products, services, and environments that foster healing, empowerment, and inclusion for all individuals, acknowledging that the past plays a significant role in shaping the present.

An example of this can be found in the *Nuka System of Care* in Alaska. This healthcare system, created for Alaska Native communities, integrates traditional healing practices with modern medical care. The design of its clinics reflects a trauma-informed approach, incorporating calming natural elements, circular seating arrangements to foster equality, and culturally relevant artwork to honor heritage. By addressing the emotional and cultural needs of the community, the design acknowledges and helps to heal intergenerational trauma.

PTSS offers a critical lens for understanding the deep-rooted tension between African American communities and law enforcement in the United States. The historical trauma of slavery, compounded by generations of systemic racism, has created an enduring legacy of mistrust and fear toward authority figures, particularly the police. This historical weight influences the behavior of many African Americans today, manifesting in nervousness or the instinct to flee when confronted by law enforcement, even in situations where they may not have committed any wrongdoing. These reactions are often driven by an inherited sense of survival, rooted in the fear that interactions with the police could escalate into danger or violence.

This dynamic, although understandable in the context of historical trauma, can sometimes lead to more unsafe situations, as heightened anxiety and fear can be misinterpreted by law enforcement as suspicious or dangerous behavior. The result is a self-perpetuating cycle where African Americans feel more unsafe around the police, and law enforcement officers, in turn, may view their reactions as threatening, further widening the gap between Black communities and law enforcement. This cycle is not just about individual interactions but also about the systemic barriers that have been built over generations, making it increasingly difficult to bridge the divide.

It's crucial to emphasize that acknowledging this dynamic is not about excusing police violence or criminal behavior but rather about understanding the complex interplay of history, trauma, and behavior. Ignoring the historical context that shapes these interactions is a missed opportunity to break this cycle. An Inclusive and Equitable Design approach could offer innovative ways to address this problem. Imagine engaging in a project with a police department where the principles of Inclusive Design are applied to create interventions that reduce fear, build trust, and foster better communication between law enforcement and African American communities. This could involve redesigning community policing strategies, training programs that address unconscious bias, or even rethinking the physical spaces where police and community members interact, all with the aim of breaking down the barriers that have been centuries in the making. By incorporating empathy and historical awareness into these designs, we could pave the way for more positive and productive relationships between law enforcement and the communities they serve.

A notable project that highlights the importance of understanding historical and cultural factors was a collaboration I had with a leading Japanese pharmaceutical company. Despite offering high-quality products and having a strong global reputation, the company struggled to gain traction in the U.S. market. We embarked on a journey with the company to identify the primary barriers it faced, particularly in terms of cultural differences. One of the key insights we uncovered was the profound impact of historical events on trust levels, particularly within African American communities.

For many in these communities, trust in healthcare organizations had been severely eroded by events that occurred generations ago, most notably the infamous Tuskegee syphilis experiment, in which Black men were used for clinical trials without their knowledge or consent. This deep-seated mistrust, rooted in historical trauma, was something our counterparts at the pharmaceutical company were largely unaware of. They had not fully grasped how such events still shaped perceptions and relationships with healthcare providers today. Once we brought this to their attention, they realized the importance of prioritizing efforts to rebuild trust and engage meaningfully with these communities rather than solely focusing on product promotion. By addressing the cultural gap and understanding the weight of history, the company was better equipped to connect with its audience and foster trust before advancing any further in its efforts. This led to revamping its marketing and messaging strategies toward clients while engaging brand ambassadors and endorsers who could advocate for the quality of the company's products and ethical practices. The company also prioritized building trust as a core value, embedding it into the company's internal culture and external communications. This focus extended to implementing routines and key performance indicators (KPIs) to track and measure progress in fostering trust across all levels of the organization.

In many discussions about inequality, particularly regarding racial, ethnic, sexual, and gender minorities, a recurring theme is the profound influence of historical and societal patterns on behavior, even when individuals consciously believe they are unaffected. A striking example of this is the pervasive experience of imposter syndrome, which tends to be more prevalent among underrepresented groups within professional or academic environments, according to several studies.[1, 2] Although various factors contribute to this phenomenon, one significant element is the way these groups are often treated—whether through microaggressions, subtle exclusion, or overly cautious or paternalistic behavior from colleagues.

These well-intentioned yet misguided behaviors—such as offering unnecessary sympathy or assuming lower capabilities—subtly reinforce the notion that members of marginalized groups are "different" or somehow require special treatment. This dynamic perpetuates the idea that their success is not entirely earned, even if this belief is never explicitly stated. These social patterns and behaviors are deeply embedded in the collective consciousness, passed down not only through generations but from one workplace interaction to another, creating a continuous cycle of doubt and exclusion.

Historical oppression, systemic discrimination, and deeply ingrained stereotypes compound these challenges, making it difficult for individuals from marginalized backgrounds to fully escape the feeling of being "othered." Whether it's racial or ethnic minorities contending with stereotypes about their work ethic or women facing subtle but constant questioning of their legitimacy in certain roles, these experiences reflect larger societal structures. These reflections, often unintentional but persistent, act as silent barriers that reinforce the unequal power dynamics within institutions, creating environments where true inclusion is difficult to achieve.

This pattern of behavior highlights how societal and historical roots, although perhaps invisible to the naked eye, continuously influence the present, perpetuating barriers that hinder true equity and belonging in professional and academic spaces.

Avoiding Stigmatization in Inclusive Design

Although the intention behind designing solutions for marginalized or underserved users can be extremely noble, the result is not always as effective as intended. More often than not, when designers create products for users who don't resemble their own experiences or backgrounds, they tend to overemphasize certain aspects, unintentionally leading to an undesired effect: stigmatization.

Stigmatization occurs when individuals or groups are marked or labeled in ways that cause them to feel ashamed, devalued, or excluded. It typically happens when a particular attribute, behavior, or identity is highlighted in a way that sets the person apart negatively from others. In design, this can manifest when products, services, or environments are created with features that inadvertently make the user feel "less than" or different from their peers, often by broadcasting their minority status or socioeconomic position. Stigmatization can be damaging because it reinforces existing stereotypes and can even contribute to social exclusion rather than inclusion.

For example, a product that is designed for low-income users and that features bold and distinctive colors might have the unintended consequence of publicly signaling the user's financial status, leading to embarrassment or feelings of inferiority. When people feel stigmatized by a product or service, they are less likely to use it, regardless of its intended benefits. This can undermine the very goal of inclusivity and lead to the failure of well-meaning initiatives.

Banks and financial institutions are prime examples when it comes to the impact of Inclusive and Equitable Design—or the lack thereof. In 2018, I was approached by a key account manager from a large scheme operator (a global payment network and credit card provider, in other words). He reached out to me urgently, asking for help.

Earlier that day, the president of a major bank, which prided itself on being "the bank for everyone," appeared live on national radio. During the interview, the journalist challenged the president on the bank's actual efforts for underbanked populations. In an attempt to showcase the bank's inclusivity, the president highlighted a few offers for people with limited incomes. But as the journalist pressed further, it became clear that the bank's efforts were insufficient. Flustered, the president hinted that new, more inclusive products and services would be revealed at an upcoming event two months later. The executive team was now under pressure to deliver, and that's how I was brought in to help develop real solutions for unbanked and underbanked people in a very short period of time.

The first step we took was to map out the existing products and understand the real needs of the target audience. The bank did have some products in place, such as credit cards specifically designed for people with low financial guarantees. These cards had limitations based on the bank's perceived risk in extending credit to this audience. However, when we asked users about their experience with these products, the feedback was harsh.

One participant shared, "This card is an insult. It only works in certain places. Just the other day, I got stuck at a highway toll because my card wasn't authorized there. It was humiliating, especially because I was driving my girlfriend.

And the design—it's bright and childish. Every time I take it out of my wallet, I know everyone around me thinks I'm poor."

This was a textbook example of how well-intentioned design can lead to stigmatization. The card, with its restricted functionality and conspicuous design, inadvertently marked users as financially disadvantaged, making them feel ashamed rather than supported. Unsurprisingly, enrollment for the product was low.

With these insights in hand, we were able to turn things around. We developed an entirely new user experience, focusing on intentional, inclusive touchpoints that respected the dignity of the users. We designed a product that met their financial needs without signaling their economic status to the world, showing that thoughtful, inclusive design can make a meaningful difference in people's lives.

By addressing the root of stigmatization and being mindful of the messages that products convey, designers can avoid reinforcing social barriers and create solutions that truly uplift their users.

Notes

[1] Bastian, R., 2019. Why imposter syndrome hits underrepresented identities harder, and how employers can help. *Forbes*, 26 November. [Online]. Available at: https://www.forbes.com/sites/rebekahbastian/2019/11/26/why-imposter-syndrome-hits-underrepresented-identities-harder-and-how-employers-can-help [Accessed: 20 January 2025].

[2] MacInnis, C. Impostor syndrome: a diversity, equity and inclusion issue. [Online]. Available at: https://www.ucalgary.ca/news/impostor-syndrome-diversity-equity-inclusion-issue [Accessed: 20 January 2025].

Strategies to Craft More Equitable Solutions

Designing equitable solutions is not just about addressing the specific needs of underserved communities; it's about uncovering opportunities to create systems, products, and environments that benefit society as a whole. The practices outlined in this chapter, such as the ladder exercise and reframing challenges, demonstrate that solutions designed with inclusivity at their core often lead to innovations that serve everyone. This principle is exemplified by curb cuts: originally designed for wheelchair users, they have proven indispensable for parents with strollers, delivery workers, cyclists, and more. Importantly, they cost no more than standard curbs to build, illustrating a crucial point: it's far more effective to anticipate barriers than to rectify them later.

The ladder exercise isn't just a valuable tool for Inclusive and Equitable Design—it's an excellent method for problem-solving in general. By zooming in on and out of a problem, designers can uncover overlooked insights and explore solutions at both the micro and macro levels. This ability to navigate different scales is central to systemic thinking, which acknowledges the interconnected nature of the challenges we face. Whether you're reimagining shampoo packaging for better accessibility or addressing global health disparities, these methods push us to think beyond surface-level fixes and consider the broader implications of our designs.

This chapter invites you to embrace strategies that reframe challenges, anticipate hidden needs, and design for impact. By demonstrating how equity-focused approaches can foster creativity and innovation, we aim to shift the discussion from one of addressing minority needs to one of universal benefit. These strategies not only enhance individual experiences but also build stronger, more resilient systems that work for everyone.

Shifting Perspectives: Reframing to Find Hidden Solutions

Once the problem has been clearly identified, the next step is to explore potential solutions and generate different options. Interestingly, the most effective way to spark ideas often begins with reframing the challenge itself. Rather than jumping straight into solutions, it's essential to further investigate the underlying issue and explore various angles from which to approach it. This reframing process can often reveal surprising insights and open up more creative, impactful solutions.

From Pills to Pots: A Case of Problem Reframing in Design

To illustrate this, let's consider a well-known case study led by Dr. Christopher Charles from the Canadian International Development Agency, who visited Cambodia for a health study.[1] During his exploration, he discovered that nearly half the Cambodian population was suffering from anemia—a condition marked by a deficiency of red blood cells or hemoglobin, which reduces the body's ability to carry oxygen. This can lead to symptoms such as fatigue, weakness, dizziness, and a weakened immune system, making individuals more vulnerable to infections.

In Western countries, anemia is typically treated with iron supplements or pills. Based on this, one could have framed the challenge as "How might we develop a food supplement to increase iron levels for this demographic?" However, by taking a step back and reframing the challenge more thoughtfully, it became clear that a different approach might yield better results.

When examining the local context, Dr. Charles and his team observed that the Cambodian diet mainly consists of fish and rice—foods that are low in iron. This insight could lead to a second framing: "How might we modify the diet of Cambodians to include more iron-rich foods?" Yet this solution presented significant challenges, including the difficulty of changing the eating habits of millions of people, not to mention the financial and logistical constraints of developing new food sources or supplements. Iron supplements may be a solution in the Western world, but in a place like Cambodia, they won't go far because of their cost, the challenge of distribution across a pretty rural country with limited infrastructure, and the relative mistrust of residents against pills. The traditional way of alleviating pain by creating an additional pill instead of working on the root cause of the problem wasn't relevant in this context.

It was key to look at the problem differently, not strictly from the symptoms angle but from the overall experience and the different places where the team could intervene.

Taking a deeper look at the problem, Dr. Charles and his team shifted focus from the diet itself to the cooking habits of the population. He realized that many Cambodians cook with aluminum pots and pans, which, although lighter than iron cookware, do not infuse food with iron. This was a key insight: using iron cookware could naturally increase iron intake without requiring a major dietary shift or expensive interventions.

From there, the challenge was reframed yet again: "How might we incorporate an iron element into the cooking process for Cambodians?"

Through further brainstorming, testing, and collaboration with local communities, the team came up with a simple yet brilliant solution: an iron fish designed to be placed in the pot during cooking (see Figure 4.1). The fish, a symbol of good luck in Cambodia, was readily accepted by the population, and it effectively infused iron into food without altering their cooking habits.

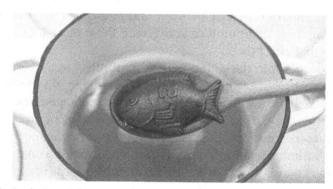

Figure 4.1: The Lucky Iron Fish

As this case shows, the turning point in the project wasn't the technology or solution itself but the ability to properly frame the challenge. By moving beyond traditional Western perspectives and exploring the problem from multiple angles, the team was able to identify a solution that was culturally appropriate, cost-effective, and highly impactful.

The Lucky Iron Fish project in Cambodia has shown promising results in addressing iron-deficiency anemia. A trial conducted in three Cambodian villages demonstrated that the iron fish ingots reduced anemia among women who used the intervention by 46% compared to a control group. Additionally, daily use of the Lucky Iron Fish has been reported to restore circulating and stored levels of iron and reduce the prevalence of anemia by approximately 43%. These findings suggest that this simple, cost-effective solution could have a significant impact on improving iron levels in populations at risk of deficiency.[2]

Framing the Challenge Helps Reduce Biases

Framing the challenge is a fundamental practice in design, and its importance is magnified in the context of Inclusive and Equitable Design.

One of the primary benefits of framing is that it allows designers to step back from the project and examine the situation from a fresh perspective. This is particularly crucial in Inclusive Design, where designers are almost never the key users of the product or service they are developing. The process helps mitigate personal biases, ensuring that the problem and potential solutions are seen through a more objective lens.

By its nature, Inclusive and Equitable Design is typically carried out by individuals who hold positions of influence or responsibility—designers, business leaders, and professionals with higher education and access to the tools of innovation. On the other hand, the subjects of their work are often people from minority groups, underserved communities, or those living with disabilities. These populations tend to have limited access to education, power, and the spheres of decision-making and creativity. This gap creates a disconnect between designers and users, which Inclusive and Equitable Design aims to bridge by empowering these communities and giving them more influence over the environments that affect them.

We'll dive deeper into designing with communities in a later chapter, and we've already discussed the role of empathy in reducing this distance, but it's important to note here that creative framing of the challenge can help designers reduce the biases they bring to the table. Although closing the gap between designers and users is essential, it's also valuable to acknowledge that designers' distance from the immediate target environment can provide a unique advantage. Sometimes users may struggle to envision solutions outside of their lived experience, as they've grown accustomed to longstanding challenges and constraints. Designers, on the other hand, can bring an outside perspective that fosters fresh, innovative thinking.

An illustrative example of user bias toward their own problem can be found in the realm of disability, specifically related to mobility challenges. Imagine a person who has used a traditional manual wheelchair for decades. Over time, they've adapted to the limitations of the chair, such as the difficulty of navigating uneven terrain, the physical strain of long distances, and the awkwardness of using the chair in narrow spaces.

When asked about potential improvements to their wheelchair, this person may not even consider a fundamentally different mobility solution. They may instead suggest incremental changes—such as improving the wheels for smoother rolling or adding cushioning for comfort—because their lived experience has made them so accustomed to the existing limitations that they can't easily envision a more revolutionary solution.

In fact, when self-driving wheelchairs and mobility exoskeletons were first introduced as possibilities, many users found the ideas far-fetched or unnecessary. After all, they had spent years finding workarounds and learning to live with the constraints of their manual wheelchairs. It took time for these new, more radical innovations to gain acceptance as users gradually realized that their own problem-solving approach had been shaped by years of adapting to suboptimal conditions.

This is a classic example of what's called *functional fixedness*, where users are so entrenched in the current way of addressing a problem that they have difficulty imagining an entirely new solution. Designers who approach the problem with fresh eyes are often better positioned to reframe the challenge and introduce innovative, transformative solutions.

The following sections explore various ways to frame your challenge to maximize creativity and ensure a thoughtful, impactful design process.

Scaling Perspectives: The Ladder Exercise for Creative Problem Solving

Imagine that you're starting your day with a normal, everyday view—the space around you, your desk, or the people in your home. Now, imagine zooming out further and further until you're not only above your home or neighborhood but far beyond the Earth itself, eventually reaching the cosmic edge of the universe. And then, just as you've taken in the vastness of space, you begin zooming back in—onto Earth, into a single person, down to their cells, and finally to the subatomic particles that make up everything around us. This mind-bending journey from the farthest galaxies to the tiniest atoms offers a fascinating perspective on scale, scope, and how systems are interconnected.

This is the experience presented in the famous video "Powers of Ten"[3] by Charles and Ray Eames. Beginning with an ordinary picnic in a park, the camera zooms out to the vast expanse of the cosmos, only to zoom back in toward the molecular and atomic levels. Similarly, the more recent "Cosmic Eye"[4] video by Dr. Danail Obreschkow takes the viewer on a visual journey from the perspective of a human eye outward to the universe and then inward to atomic levels. Both videos highlight how vastly different things can appear when you change your perspective—looking at the same object from a macro and a micro level.

These videos perfectly illustrate the ladder exercise I learned from Clark Kellogg at UC Berkeley. Clark taught Design Thinking at Berkeley for over two decades, and we collaborated on many programs. The exercise is a powerful tool in design thinking that teaches us to shift our perspective by zooming in and out of a problem; just as the videos show us the scale of the universe, this exercise helps us explore the challenges we face on multiple levels.

By taking this approach, much like in the videos, you gain fresh insights and uncover solutions that may not have been obvious before. The videos remind us that the way we frame a problem and the perspective we choose to adopt can greatly influence our understanding and the solutions we develop. Whether you're solving a design problem, addressing a social issue, or tackling a technical challenge, changing your perspective—zooming in and out—can make all the difference in arriving at a truly impactful solution.

The ladder exercise is pretty simple. Start with your current challenge, and try to zoom out several times to look at it from a wider perspective, taking into consideration the system in which the challenge exists. Then go back to the beginning and zoom in: look at the details . . . the smallest part of the problem . . . the grain of salt that makes something go wrong.

Doing this helps change your perspective on the challenge to explore zones and objectives that were not anticipated earlier.

To illustrate, consider the case of the personal care brand mentioned in Chapter 3, "Systempathy—or How to Empathize with People and the System They Live In." My client and I wanted to consider how to make shampoo more inclusive and equitable. Zooming out, we looked at how we might use the beauty industry to create a more gender-inclusive society. We examined the whole beauty industry and questioned its role in perpetuating divergences between genders. We considered the entire system and the incredible reach of the beauty industry in terms of products and marketing.

Then, moving down the ladder, we looked at how we might make the text more visible on the shampoo bottle. This led us to focus on another form of exclusion, particularly for individuals with visual impairments. It even included people with mild hyperopia, who often struggle to read small or unclear labels, especially in environments like the shower. The risk of mixing up shampoo and conditioner due to unreadable labels is an everyday inconvenience for many people, so we explored ways to make packaging more accessible, ensuring that product use is intuitive and inclusive for all users.

Another highly effective version of the ladder exercise is to shift the focus from perspectives to timeframes. This method allows us to explore solutions based on varying time horizons, prompting us to think about immediate, short-term, medium-term, and long-term impacts. By considering what can be accomplished in 3 hours, 3 days, 3 months, 3 years, and even 30 years, we gain a multidimensional view of the challenge and its potential solutions.

Let's apply this approach to our challenge: how might we make shampoo more inclusive?

We'll tackle the challenge in:

3 Hours

In a 3-hour window, the solution must be quick and impactful, using minimal resources. One potential step could be making minor but significant packaging adjustments: for instance, adding braille labels to shampoo bottles to assist visually impaired users or attaching a simple QR code that links to audio instructions or information. Another solution could involve a swift social media campaign offering more diverse hair care tips representing different hair types and textures.

3 Days

With 3 days to work with, we could focus on launching a small-scale pilot program in a local community to gather user feedback. This could involve offering trial-size shampoos in personalized packaging that reflects diverse hair types or creating scent-free options to cater to users with sensitivities or allergies. Partnering with community organizations, we could also host an awareness event on inclusivity in beauty and hair care, creating immediate traction and dialogue.

3 Months

In 3 months, a more comprehensive solution can be developed. One option could be creating a line of inclusive shampoos that specifically cater to under-represented communities, such as individuals with different hair textures or those from socioeconomically disadvantaged groups. This could include collaborating with diverse hair experts, conducting focus groups, and designing a nationwide marketing campaign targeting inclusivity in hair care. Additionally, educational initiatives could be introduced in-store or online to raise awareness about inclusive beauty.

3 Years

With 3 years, we have the time to implement systemic changes within the brand and company. We could invest in Research & Development (R&D) to innovate formulas that cater to all hair types and textures, ensuring a truly universal product. This could also be the timeline for initiating a full rebranding campaign where inclusivity is not just an addition but the core identity of the company. Partnerships with diversity, equity, and inclusion (DEI) experts, creating mentorship programs for minority beauty entrepreneurs, and leveraging technology such as AI personalization tools could also be implemented to ensure that every customer feels represented.

30 Years

Over the span of 30 years, the long-term impact of Inclusive Design could involve global industry shifts in the way beauty is perceived and marketed. The shampoo brand could lead the way in championing sustainable and equitable production processes, ensuring that every part of the supply chain—from ingredient sourcing to manufacturing and retail—supports inclusivity and sustainability. We could also aim to standardize Inclusive Design practices across the beauty industry, advocating for inclusivity in product development, marketing, and workplace culture.

This exercise showcases how looking at the same challenge through different timeframes allows for a broader range of creative and impactful solutions, from immediate fixes to systemic transformation.

The ladder exercise is not only a powerful tool for Inclusive and Equitable Design but also a highly effective approach to problem-solving across various contexts. Whether you're designing a product, developing a service, or tackling any complex challenge, this exercise helps you see the bigger picture while also uncovering overlooked details. By consistently zooming in and out, you can gain a deeper understanding of how various elements interact within a system and how those elements influence the outcome of a solution.

The broader utility of this exercise lies in its ability to force you out of tunnel vision. Many times, problem-solving gets bogged down in a narrow focus on immediate issues without considering the larger context or the underlying root causes. The ladder exercise encourages you to step back, allowing for a broader systems-thinking approach that considers external forces, interdependencies, and cultural or societal impacts. Then, by zooming in, you sharpen your focus on the specific pain points or micro-level details that can make or break a solution.

In the context of Inclusive and Equitable Design, this dual perspective is not just helpful—it's essential. Our aim is to create solutions that account for and respect the diversity of human experience, and that means understanding both the broader system in which a problem exists and the individual experiences within that system. This ability to see the entire ecosystem while also understanding the finer, more personal details of users' interactions allows for more thoughtful, comprehensive, and impactful solutions.

In other words, although the ladder exercise is applicable across many problem-solving scenarios, it becomes a requirement for Inclusive and Equitable Design because it enables us to see beyond the immediate challenge. It helps us grasp the system we're working within—whether that's a societal structure, an industry dynamic, or a cultural context—and ensures that our solutions are not only relevant to the task at hand but also sustainable and transformative for the larger system we are trying to influence.

From Milkshakes to Remittances: Applying "Jobs to Be Done" in Design

Another powerful method to expand our range of solutions is by applying Clayton Christensen's *Jobs to Be Done* theory, a well-known innovation framework. This approach shifts our perspective from focusing solely on product features to understanding the fundamental "job" that customers hire a product or service to perform.[5]

The Jobs to Be Done framework emphasizes that people don't simply buy products; they "hire" them to perform a specific job. To illustrate this, Christensen used a famous example involving milkshakes at a fast-food restaurant.

In this case, a restaurant wanted to increase milkshake sales but found that traditional customer surveys weren't providing clear answers. Instead of asking customers how to improve the milkshake, Christensen's team asked a different question: what "job" were people hiring this milkshake to do?

Through observation, the team discovered that many people purchased milkshakes in the morning before long commutes. These customers wanted something that would keep them entertained, be easy to drink in the car, and provide enough sustenance to last until lunch. Interestingly, it wasn't just about taste or ingredients—it was about the milkshake fulfilling a specific role in their routine.

With this insight, the company didn't need to change the flavor of the milkshake but rather focused on improving its thickness and packaging, ensuring that it took longer to finish and remained satisfying during long drives.

By focusing on the job customers were trying to achieve rather than just the product itself, the company could tailor its offering to better meet customers' real needs.

This approach can be incredibly valuable in Inclusive and Equitable Design, as it helps us understand not only the products we are designing but also the deeper needs and jobs that users are trying to fulfill, especially when designing for underserved communities or people with disabilities.

This approach is particularly useful for understanding the current alternatives and coping mechanisms that users are employing to solve their problems. In the case of milkshakes, for example, the alternatives aren't limited to other milkshakes but also include bananas, bagels, energy bars, and other snacks. By focusing on the "job to be done"—something that would keep them entertained, be easy to consume in the car, and provide enough sustenance to last until lunch—Christensen realized that bananas are finished too quickly, bagels are so dry that they often require cream cheese (which can be messy while driving), and energy bars, although convenient, aren't always the healthiest option.

People with disabilities exemplify this kind of creative problem-solving, often using limited resources to get the job done. As I write this, the Paris Paralympic Games are wrapping up, and each event showcases inventive adaptations of mainstream sports. One striking example is blind football, where the ball is fitted with bells to make sound as it moves, allowing players to track it by ear. During penalty shootouts, coaches tap the goalposts with a steel rod to help players orient themselves solely through sound before taking their shot.

This is just one example of how creativity flourishes in adapting to unique challenges. There are countless others, such as these:

■ **Blind individuals using ropes and bells:** In homes or environments where accessibility features are limited, blind individuals sometimes set up physical cues such as ropes or bells to help navigate spaces. For instance, ropes attached to walls may guide someone to different rooms, while bells attached to doors or objects provide auditory cues for location. This technique is highly effective in places where modern accessibility technologies, such as smart home systems, are unavailable.

- **Deaf individuals using vibrating alarms and light signals:** In homes, many deaf people use alarms and alert systems that are designed to vibrate or flash lights. For example, instead of a traditional doorbell, some homes feature doorbells that cause lights to flash in a specific pattern when pressed. Similarly, vibrating alarms placed under mattresses can wake deaf individuals without the need for auditory cues.

- **Rubber bands for tactile marking:** People with visual impairments sometimes wrap rubber bands around certain items—like shampoo and conditioner bottles—to differentiate between them by touch. This method helps ensure that they use the correct product, especially in situations where different bottles look and feel the same otherwise.

- **Different textured materials for orientation:** In homes or workplaces, individuals with visual impairments often place mats or textured floor surfaces in specific areas. For example, using a rough doormat before a bathroom can indicate where a person is standing. These tactile indicators can guide someone throughout a space without the need for more expensive or complex adaptations.

- **Mirror positioning for mobility in small spaces:** In narrow or cluttered environments, individuals with physical disabilities may place mirrors at strategic angles to see around corners without needing to physically turn around. This allows them to navigate spaces more easily, especially when using a wheelchair.

All of these examples show that solutions don't need to stay within the boundaries of existing and mainstream options but can instead be more creative in their approach to the overall environment.

Similarly, when working on design challenges, it's crucial to look beyond direct competitors and examine how people are finding alternatives to solve immediate problems. For instance, in a research project where I supervised a collaboration between my students and one of the largest blockchain startups in the world, with a focus on blockchain solutions for remittances, instead of only analyzing other currencies or banking mechanisms, we investigated the informal and creative alternatives that people use to send money across borders.

For example, foreign workers often rely on the hawala system, an ancient and informal money transfer network that operates on trust. Instead of physically moving cash across borders, brokers facilitate the exchange of funds between countries using a balance of debts. This system can bypass official financial channels but also poses challenges related to transparency and legality.

Another common practice is using courier services or trusted individuals to carry cash across borders physically. Although this method can be risky due to theft or legal concerns, it remains a viable option for many. The potential for loss or delays adds a layer of complexity, but it is a lifeline for individuals without formal banking access.

Finally, gift cards or vouchers are sometimes used as an informal remittance method. A sender purchases a gift card for a global brand or retailer and sends the card to a family member, who can then use it locally. This bypasses the need for formal banking while still providing value, although it limits how the recipient can use the money.

By examining these alternatives, we better understand the real needs and limitations faced by users. This insight helps us identify the key factors to focus on during the design process, much as Christensen discovered that the key attributes for improving milkshakes were related not to flavor but rather to thickness and packaging. In remittance systems, key factors may include speed, security, trust, and accessibility, which can then guide the ideation process toward solutions that effectively meet users' needs.

For example, the hawala system highlights the importance of trust, and the courier system emphasizes the need for secure transport. Similarly, the rise of cryptocurrencies speaks to the desire for decentralization and bypassing formal regulations, whereas gift cards illustrate a creative workaround for avoiding fees or currency restrictions.

Looking at these alternatives not only informs us about the creative methods people use to solve their problems but also helps refine the specifications of what we need to design, leading to more tailored and effective solutions.

Being Creative—A Pragmatic Exercise

Although creativity is often perceived as an innate gift—something that some people are simply born with—the reality is that anyone, whether an individual or a team, can unlock their creative potential through structured processes and deliberate steps. Creativity doesn't always mean breaking the rules or venturing into the unknown; in my experience, it can be a highly pragmatic exercise grounded in thorough exploration and the ability to transfer ideas from one field to another.

The techniques and approaches that follow have been around for some time and are widely used in innovation and design projects beyond Inclusive Design. However, they have proven particularly effective in the context of Inclusive and Equitable Design, offering valuable tools for leaders, product managers, designers, and anyone passionate about both Design and DEI. By applying these methods, organizations can foster more thoughtful, impactful, and accessible solutions that truly serve diverse communities.

Finding Solutions in Plain Sight: A Side-Step Approach to Inclusion

In 2021, I found myself working with a globally renowned wine and champagne producer, specifically their team based in Napa Valley. Our connection began through a professional network, and soon I was having lunch with the CEO of their US subsidiary. As we sat on the beautiful grounds of the vineyard, I took the opportunity to explain my work on Inclusive and Equitable Design, curious to see if any of it resonated with his experiences. He paused thoughtfully and then shared an unexpected insight. The local labor pool was strikingly homogeneous. Being in a region so centered around the wine industry, the location itself presented a challenge for diversification. The area's small, scattered cities and lack of public transportation made it difficult for workers to commute unless they lived nearby. There was a tourist train, but that wasn't exactly a solution for the workforce.

Despite these structural barriers, the CEO expressed pride in the racial diversity they had managed to cultivate within the team. However, he admitted that integrating people with disabilities was an area where they had fallen short. It wasn't for lack of desire—he genuinely viewed it as a critical topic—but he struggled to attract employees with disabilities given the region's challenges. That conversation became the spark for what would turn into a collaboration that drove some significant and innovative changes within the company.

Our first task was to examine the various roles and jobs within the company to better understand what barriers existed for people with disabilities and identify which roles might be the best fit. As we began our audit with the HR team, we decided to take a different approach. Rather than imposing our assumptions about what roles might or might not be accessible, we flipped the script. We asked them, "What are the most critical jobs in your company right now?"

They immediately mentioned the prestigious role of winemaker—understandable, given that this role is like the artist behind the masterpiece. But something unexpected came up during the conversation. One HR team member brought up a job opening they were struggling to fill: a bottling line operator. It was a relatively low-qualification role, with good pay, responsibilities, and flexibility, but it was plagued by high turnover. People weren't staying in the position for more than a few weeks or months, and the company was desperate to find a solution.

Intrigued, we dug deeper. We requested a site visit to observe the production line firsthand, and the problem became clear almost immediately. The work environment in the warehouse was overwhelming—particularly the noise. Large machines ran constantly, creating an unbearable level of sound. Although workers were provided with protective equipment, it was still physically and mentally exhausting to endure the incessant noise, which often left them with a persistent ringing in their ears at the end of the day. No wonder they didn't stick around.

At that moment, we shifted our perspective yet again. Instead of trying to adapt existing jobs to people with disabilities, we asked ourselves, what if we considered people with disabilities not as limited but as individuals with alternative abilities? We realized that someone with an auditory impairment or even someone who was deaf would be perfectly suited for this bottling line role. They wouldn't suffer from the noise in the same way a hearing person would, and in fact, they might even excel in the position. This was an untapped opportunity.

This experience taught us an important lesson about how easily we can find creative, relevant solutions when we embrace the differences held by those who are traditionally excluded. Instead of viewing disabilities as barriers, we began to see them as potential advantages in certain contexts. The key was in matching the unique attributes of individuals to the specific challenges our client was facing.

This story also reminded me of an earlier case involving a real estate broker from an underserved area whose deep knowledge of local neighborhoods and proximity to highways allowed him to save his company during the pandemic. His network, often overlooked by mainstream businesses, proved to be a game-changer, much as individuals with different abilities can bring fresh solutions to old problems when given the chance.

In both cases, the solutions were hiding in plain sight. It just required taking a side step and reframing the challenge to see the possibilities. And in both cases, the results were an immediate return on investment with the ability for the company to maintain its activity and profitability and also a deep change in the team's perspective on the importance of diversity. With those immediate examples, it became very clear to people how much they owe those colleagues in the delivery of their mission and, by extension, their own incentives.

This example of rethinking workforce diversity in a champagne producer's operations offers a powerful demonstration of how Inclusive Design can lead to creative, practical solutions by reframing what we perceive as limitations into opportunities. Rather than viewing disabilities as barriers, this approach encourages us to explore how the unique attributes of individuals, especially those who are often marginalized, can be leveraged to solve specific challenges.

With this in mind, we also want to make sure we are not using this as an excuse for companies to avoid investing in improving the workplace. As stated already, the ultimate goal is to create solutions that benefit everyone, including those who are usually excluded, not to use people with specific abilities as a bandage to cover a problem that should be fixed for the greater good.

This innovative approach can be applied to broader areas of Inclusive and Equitable Design, extending beyond workforce diversity. For instance, in product design, similar reframing can help make products more accessible to a wider audience. Consider smartphone accessibility features: instead of viewing disabilities as challenges to overcome, companies like Apple have designed products that incorporate voice-over functionality, text-to-speech, and voice commands. These features were initially developed for people with visual impairments or mobility issues, but they have since become widely adopted by all users because of their convenience and flexibility.

In urban planning, this same principle can guide efforts to make cities more accessible. For example, curb cuts, originally designed for wheelchair users, have proven to benefit a much wider range of people, including parents with strollers, delivery workers, and travelers with luggage. The initial problem—how to make cities more accessible for disabled individuals—was reframed to uncover a universal solution that serves a broader population.

Similarly, in education, reframing the challenge of meeting diverse learning needs has led to innovations like universal design for learning (UDL). Instead of seeing students with learning differences as exceptions that require special accommodations, UDL principles emphasize designing curricula and teaching methods that are flexible enough to accommodate all learners from the start, creating a more inclusive and equitable educational environment.

These examples demonstrate that when we stop trying to "fix" differences and instead find ways to leverage them, we can uncover innovative solutions that benefit everyone. Inclusive and Equitable Design is not about catering to the few; it's about creating systems, products, and environments that are adaptable, accessible, and enriching for all.

Borrowing the Best: How Cross-Industry Solutions Elevate Inclusive Design

When it comes to solving problems, especially in the realm of innovation, it's tempting to chase after groundbreaking, never-seen-before solutions. As innovators, we often aspire to be pioneers, eager to create something entirely new. The pull is even stronger when working in the field of Inclusive and Equitable Design. The purpose is noble—it's about making the world more accessible and fair—so there's this extra urge to be the hero, the one who invents something revolutionary to solve a critical issue.

But in my experience, the best approach is often to resist that urge. Instead, I always advise my teams and students to first take a step back and explore existing solutions. We don't need to reinvent the wheel; there are countless efficient, proven ideas in other domains that can be adapted to our specific challenges. This analogy work—where we borrow solutions from one industry and apply them to another—has consistently proven to be effective. It reduces the risk of failure because you're leveraging solutions that have already been tested. It speeds up the process by allowing you to build on the successes of others. And, crucially, it can save time when pitching new concepts because stakeholders may already be familiar with these ideas from other contexts.

One project that perfectly illustrates this principle is our recent collaboration with the US leader in notebook manufacturing. I assembled a team of young and talented designers and consultants and led a research effort with key strategic recommendations, product releases, and improvements to support a more Inclusive Design. Like many of our partnerships, it started with a well-intentioned goal. designing notebooks for everyone. However, the plan was rather vague at first, with no clear direction. Through several discussions, I helped the team narrow their focus. We explored various underrepresented audiences and eventually decided to focus on people with neurodivergence as our primary demographic.

The company wasn't just producing physical notebooks; they were also developing a complementary digital app. Our mission became twofold: to make both the physical product and the digital experience more inclusive for neurodivergent users. However, we had to operate within the constraints of a large organization—economic considerations, industrial limitations, and the company's well-established reputation. These limitations made the task more challenging, but they also fueled our creativity and determination.

Once we framed the challenge and conducted our research, we decided to do exactly what I always recommend: look at existing solutions from other industries. Instead of trying to invent something entirely new, we explored ideas that were already working in different contexts. We identified tools and strategies from fields like assistive technology, gaming, and therapy and adapted them to fit the unique needs of this project. By doing this, we were able to deliver solutions that were not only innovative but also practical and ready for real-world implementation. This practice is sometimes referred to as *analogous observations*.

Here are four key solutions we came up with that we borrowed from other industries and contexts:

Collaboration and study group feature   The idea of incorporating a real-time collaboration feature in the study app was inspired by the use of collaborative workspaces commonly seen in the tech industry (e.g., tools like Google Docs and Slack). In industries where teamwork is essential, such as software development and project management, real-time collaboration tools have been instrumental in boosting productivity and engagement. By adapting this concept for the study app, we aimed to foster community and engagement among both neurotypical and neurodivergent students, allowing them to form supportive study groups and share notes in real time. This solution brought a sense of community to the education space, making it easier for students to connect and engage while reducing feelings of isolation often experienced by neurodivergent individuals.

Reward system for gamified learning   The reward mechanism integrated into the study app, which includes badges and virtual currency, was borrowed from gamification techniques used extensively in the gaming and fitness industries. These techniques are widely known to boost engagement and motivation by providing immediate feedback and rewards for progress. Fitness apps, for example, use badges and points to encourage users to stay consistent with their workouts. This concept was applied to education, where neurodivergent students often face difficulties with sustained focus and motivation. By rewarding achievements such as completing study sessions or mastering a topic, the app incentivized learning and kept students engaged—much as a fitness app would motivate users to stay active.

Text-to-audio conversion   The text-to-audio feature, which allows users to convert written text into audible content, was inspired by assistive technologies originally designed for visually impaired individuals. This technology is commonly used in screen readers for web accessibility, enabling those with visual impairments to access written content through audio. By adapting this solution for neurodivergent students, the project provided an alternative way for individuals with dyslexia or Attention-Deficit/Hyperactivity Disorder (ADHD) to consume information, catering to diverse learning styles. This is a perfect example of how a solution developed for one type of disability was transferred to another context, broadening its impact.

Tactile notebook covers   Another successful transfer of an idea was the introduction of sensory-stimulating notebook covers (see Figure 4.2). Tactile elements, often used in therapeutic tools for individuals with autism or ADHD, help regulate sensory input and provide calming effects. These are common in sensory therapy products such as stress balls and fidget toys. By incorporating this sensory-friendly design into everyday school supplies like notebooks, we were able to help students stay focused and manage their sensory sensitivities in a classroom setting.

Figure 4.2: Mockup of the tactile notebook cover made with Dall-E

This project is a testament to the power of borrowing ideas across industries. When we abandon the notion that every solution must be original and instead look to proven solutions from other fields, we unlock opportunities to make rapid, meaningful progress. In this case, solutions from the tech, gaming, and therapeutic industries transformed the way we approached Inclusive Design for neurodivergent students. By drawing from existing ideas, we not only saved time and reduced risks but also created a framework that was immediately relatable to stakeholders. In doing so, we proved that innovation isn't always about creating something entirely new—it's about adapting what works and making it relevant to your specific context.

Four Ways to Make a Difference—The Solution Matrix

The Solution Matrix, as shown in Figure 4.3, was born out of a need to help students push the boundaries of their thinking when working on their Equitable Design projects with corporate partners. Early on, I noticed that when students were tasked with proposing solutions, they often defaulted to safer, more traditional ideas—those aligned with their partner's existing business models or ways of thinking. This wasn't necessarily bad, but it limited creativity and often led to incremental changes rather than bold, innovative solutions. So, I created an inclusivity-oriented variant of a traditional matrix—the Solution Matrix—to encourage students to think more broadly and challenge their corporate partners in a structured way.

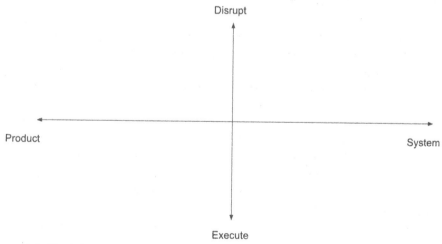

Figure 4.3: The Solution Matrix

The matrix is built on two key axes: the Execute–Disrupt axis and the Product–System axis. The Execute–Disrupt axis is about how solutions are framed. On the Execute side, students are encouraged to align with their partner's existing way of thinking. This means they work within the current constraints, improving or enhancing what already exists. On the other end of the spectrum is Disrupt, where students are asked to challenge the status quo. Here, they are encouraged to think radically and propose solutions that go against the grain of their partner's typical approach.

The second axis is the Product–System axis. This is about what the solution impacts. On one side is Product, where students focus on developing new or improved versions of existing products. On the other side is System, where they look at solutions that impact the broader organizational or industry ecosystem, often requiring more systemic changes.

The magic of the Solution Matrix lies in how it forces designers to fill all four quadrants: Product–Execute, Product–Disrupt, System–Execute, and System–Disrupt. By doing this, designers aren't allowed to settle on a single type of solution. They are pushed to think both conservatively and radically, at both the product and system levels. This structure not only ensures a diversity of ideas but also increases the likelihood that at least one of the proposed solutions will resonate with the partner or sponsor.

Let me give you an example. Imagine a company looking to address DEI within its recruitment practices. Applying the Solution Matrix, the Product–Execute quadrant might involve an incremental change, such as updating job descriptions to include more inclusive language and using software that checks for gender bias. This would be a direct, actionable solution that improves an existing aspect of the process.

In the Product–Disrupt quadrant, designers might propose a more radical change, such as the creation of a completely new recruitment platform designed specifically for diverse candidates, incorporating AI to match not just skills but also cultural fit, neurodiversity, and life experiences. This would challenge how the company currently recruits and force the company to consider a new approach.

Moving to the System–Execute quadrant, designers could look at enhancing the company's internal culture and policies to make the workplace more inclusive for all employees. This might involve setting up mentorship programs, expanding DEI training, or creating affinity groups. These solutions would work within the existing framework of the company but take a broader view of systemic changes that can be made.

Finally, in the System–Disrupt quadrant, designers might propose something far more transformative, such as the company leading an industry-wide coalition to establish new standards for DEI hiring practices or lobbying for legislation that mandates diversity quotas. This approach would push the company beyond its own operations and into a leadership role within its industry, completely disrupting how DEI is handled on a systemic level.

By structuring solutions this way, the Solution Matrix pushes teams to be creative, bold, and strategic, providing their partners with a range of options from safe to radical. The matrix doesn't just guide ideation—it enhances the likelihood of implementation. Corporate partners can see a full spectrum of possibilities, and even if they aren't ready to leap into a disruptive solution, they may still implement an execute-level idea or a smaller product innovation. In the end, this method fosters a richer dialogue between designers and partners or clients, encouraging both innovation and practical application.

Courageous Conversations: Unlocking Equity in Design and Innovation

In the world of Inclusive and Equitable Design, challenging the status quo often begins with having the courage to engage in uncomfortable conversations. Whether it's addressing systemic inequities, questioning ingrained biases, or advocating for underrepresented voices, these dialogues are critical to designing solutions that foster inclusion and justice. The following sections explore how these tough conversations—both within teams and with clients—have the power to unlock creative and bold approaches to equity in design and innovation.

Uncomfortable Conversation with an Equitable Designer

It takes an incredible amount of courage and determination to engage in uncomfortable conversations—particularly when those discussions revolve around race, equity, and the deeply ingrained systems of inequality in our society. One person who has embraced this challenge and made a profound impact is Emmanuel Acho, the man behind the popular series "Uncomfortable Conversations with a Black Man." You may have come across his videos, read his book, or heard him speak on platforms about race and social justice.

Emmanuel Acho's work is remarkable because he demystifies these difficult conversations, offering a space for both his guests and his audience to confront topics that many are hesitant to address. He doesn't shy away from the tough questions. Instead, he encourages people to ask them, to embrace their discomfort, and to listen with empathy. Acho's tone is neither accusatory nor defensive—he creates an environment where ignorance can be addressed without judgment and where people feel safe to explore their own biases and misconceptions. His conversations provide a bridge for those who might otherwise avoid these topics out of fear or discomfort.

For instance, in one of his episodes, Acho sits down with Matthew McConaughey to discuss White privilege. The conversation is honest and raw, with McConaughey asking candid questions about his role as a White man in a racially biased society. Acho's response is compassionate yet direct, explaining how systemic racism works and how individuals can recognize their unearned advantages. By leading with understanding and offering clear, actionable insights, Acho makes these uncomfortable conversations educational without being confrontational, offering his audience a pathway to engage with social issues in a more meaningful way.

Where Emmanuel Acho's work is deeply relevant to Inclusive and Equitable Design is in the practice of bringing these difficult conversations into the design process itself. In traditional design thinking, we often focus on user needs, functionality, and aesthetics, but Inclusive Design takes it a step further by actively questioning which voices are missing from the table and addressing the inequities built into many products and systems. It requires designers to confront biases—both their own and those embedded in the systems they are designing for. This can lead to moments of discomfort when designers must ask themselves, "Am I unintentionally excluding someone?" or "How might this design perpetuate inequities?"

As in Acho's conversations, there are moments in the design process when questions arise but the fear of appearing ignorant or being perceived as offensive can paralyze even the most well-intentioned designer. This hesitation can lead to missed opportunities to create more inclusive solutions. For example, a designer working on a new public transit system might avoid asking whether it properly serves low-income communities, worried that the question would be seen as patronizing or controversial. Or, during a product design meeting, someone might hesitate to question why certain accessibility features, such as voice commands or tactile interfaces, aren't included—afraid of being labeled as uninformed about disability rights. These moments of avoidance can stifle innovation and result in products that perpetuate exclusion.

This is where Equitable Designers step in. They not only initiate these tough conversations but also build them into every step of the design process, ensuring that the uncomfortable realities of inequity are confronted head-on. By doing so, they open entirely new horizons for solving the deeply ingrained issues of exclusion and discrimination. For instance, consider a design team working on a financial literacy app. Without engaging in uncomfortable conversations about socioeconomic disparity and access, they may overlook the fact that many low-income users do not have consistent Internet access. But by addressing these realities—by asking the uncomfortable questions—they can design a product that works offline or requires minimal data usage, thus making financial education more accessible.

Just as Emmanuel Acho makes it possible for people to cross the chasm and engage with difficult issues around race, Equitable Designers create frameworks where uncomfortable questions can be asked, biases can be confronted, and new, inclusive solutions can be discovered. The act of designing becomes not just about creating a product or system but about creating opportunities for equity, for inclusion, and for a better world. Through these courageous conversations, both the designer and the end user are empowered to shape a more inclusive and just future.

Brave Conversations for Lasting Change: A Partnership to Empower Black Founders

One of the most transformative experiences of my career involved a project with a world-leading consulting firm and its Corporate Venture Capital arm. Acting as consultants with a team of students I supervised, our goal was ambitious: to bridge the resource gap for Black founders, who receive only 0.5%[6] of venture capital funding. We focused on universities as an essential piece of the puzzle and how to improve the resources in this environment. We knew from the start that this would require uncomfortable conversations—not only with the stakeholders but also within our own team. These weren't just the kind of conversations where we exchanged ideas about market gaps or investment strategies. These were deeper, more raw discussions about systemic inequities, implicit biases, and the structures that perpetuate exclusion.

From the outset, we were faced with the stark reality that 88% of Black founders were self-funded and 90% found university resources difficult to access.[7] Despite the good intentions of universities and accelerators, the resources simply weren't reaching the people who needed them the most. And then there was the most glaring statistic: 100% of the founders mentioned the need for financial stability, a need that had gone largely ignored by existing programs. These numbers forced us to confront some hard truths—not just about the ecosystem but about how we ourselves had been engaging with it.

We began by meeting with our partners at the consulting firm and their Corporate Venture Capital arm. It was clear that they wanted to make an impact, but there was hesitation in how far they were willing to go. They had launched a Black Founders Development Program (BFDP), which was a great start, but as we dug deeper, it became clear that they were operating within a comfortable framework—one that didn't fully address the structural challenges Black founders face. This is where the uncomfortable conversations began.

In one of the first strategy sessions, I remember raising the issue of mentorship. Most Black founders lacked access to industry-specific mentorship, but mentorship alone wasn't enough. They needed long-term financial security, and often they weren't getting that from the programs offered by large organizations. I could see the tension in the room. One executive from the consulting firm asked, "Do you really think we can tackle financial inequity through a mentorship program? Isn't this beyond our scope?" It was a valid concern, but the question itself pointed to a larger issue: we were thinking too small.

That was the moment when the conversation shifted. We had to move away from what was easy and comfortable and toward what was necessary. The real problem wasn't just access to mentorship—it was the inequitable distribution of resources at a systemic level. We needed a solution that went beyond the surface. We proposed two solutions that would stretch the boundaries of what their BFDP had initially intended: a Steward Search Tool to increase accessibility to university resources and a financial stability framework that would ensure these founders had guaranteed jobs either at the consulting firm or with industry partners.

The Steward Search Tool was a breakthrough. It allowed Black founders to connect directly with mentors and alumni in their industry, breaking down the barriers to mentorship that had long been a roadblock. But more than that, it also created a system that was easy to navigate, increasing the likelihood that founders would find the resources they needed. We had designed a tool that made accessing mentorship simple, but we knew that mentorship without financial backing wouldn't be enough to address the systemic inequities these founders faced.

This is why the second part of the solution—a financial stability framework—was critical. We didn't want to merely guide Black founders to potential opportunities; we wanted to guarantee them jobs with the consulting firm or through partner organizations. This was a tough sell. The idea of guaranteeing jobs or financial support was met with resistance. Some questioned whether this level of commitment was feasible or even aligned with the firm's objectives. This was where the real courage was required—both on our part and on the part of our partners. It wasn't easy to advocate for such a bold approach, but we pushed forward.

Amid all this, there was an added layer of complexity on our own team. Only one team member was Black—a man with extensive experience working in startups. His knowledge and personal connection to the issues at hand made him an invaluable part of the group but also created some discomfort within the team. Many of us felt a fear of not being legitimate in these conversations about race and systemic inequities. The unspoken hesitation—are we even qualified to ask these questions?—created an atmosphere of restraint. People held back from fully engaging because they didn't want to seem ignorant or offensive in front of him.

This discomfort could have derailed the project, but instead it became the turning point in our journey of Equitable Design. We recognized that if we wanted to create real change, we needed to start by creating a safe and brave space within our team: safe, so that people could ask difficult questions without fear of judgment; and brave, because we had to be willing to confront our own insecurities and biases. Our Black colleague encouraged us to engage openly, reassuring us that although he had lived experience, he wasn't the sole voice of Black founders.

Once we began to have more candid conversations internally, it was as though a floodgate opened. The team started generating ideas without fear, and we were able to bring that same openness into our conversations with the client. We were no longer constrained by fear of stepping out of bounds; instead, we were pushing for the bold solutions this project demanded. The discussions became more honest and more impactful, which allowed us to address the systemic issues more comprehensively.

In the end, we moved forward with both solutions. The Steward Search Tool would make university resources more accessible to Black founders, and the financial stability framework would provide concrete opportunities for long-term success. The consulting firm's Corporate Venture Capital arm recognized that by doing this, they were not just helping Black founders—they were reshaping the entire ecosystem.

This project wasn't only about the solutions we delivered; it was about the courage to have uncomfortable conversations, challenge the status quo, and push for real change. By stepping into that discomfort, we created an environment where ideas could flow freely, and we came together as a team. Ultimately, we crafted solutions that not only addressed the immediate needs of Black founders but also tackled the deeper systemic inequities. It was a powerful reminder that true innovation in Equitable Design is born from having the courage to confront uncomfortable realities—and doing so together as a united team.

Breaking Taboos: Navigating Sensitive Conversations in FemTech Data Security

In the same capacity, I supervised a team of students working on a female tech project as we tackled a unique challenge centered around the privacy and security of sensitive data in the FemTech space, particularly related to women's health and menstrual cycles. Initially, this project introduced an unexpected layer of difficulty—not in terms of technology, but in the conversations we had to have. Discussing topics like menstruation and feminine hygiene in a professional setting, particularly on a team composed of individuals from diverse genders, cultural backgrounds, and religious beliefs, made many team members uncomfortable. There was an unspoken barrier when it came to diving deep into these subjects, given the various taboos surrounding them in different cultural and personal contexts.

The political landscape in the United States only heightened the sensitivity of the topic. The overturn of Roe v. Wade had placed women's health data at risk, making it more vulnerable than ever. Many women were using apps to track their periods, but they weren't aware that their personal data—information as intimate as their menstrual cycles—was being shared with third-party companies without their consent. This meant we weren't just dealing with a technical issue; we were navigating the complexities of data security in an era of heightened political and personal stakes.

When we first sat down as a team to discuss the project, there was hesitation. Some of the male team members, for example, seemed unsure of how to engage in conversations about periods or women's health without overstepping boundaries. Others, coming from more conservative cultural backgrounds, found it difficult to openly discuss a subject that had long been considered taboo in their communities. This discomfort manifested in polite but surface-level conversations that often skirted the core of the issue: women's vulnerability around sensitive health data.

But we quickly realized that if we were going to create a meaningful solution, we needed to push through that discomfort. Instead of avoiding the topic, we began addressing it head-on. We consciously created a supportive space within our team where every question could be asked, no matter how awkward it might seem. This meant encouraging open discussions about the fears, risks, and data privacy concerns that come with being a user of these apps and confronting the gender biases that often make topics like menstrual health difficult to discuss in professional settings.

As these conversations evolved, so did our understanding of the problem. We realized that the core issue wasn't just technical—it was also about cultural attitudes toward feminine hygiene. Women's health and the data surrounding it had long been neglected or trivialized, even though this information is deeply personal and politically charged, especially in regions with strict reproductive laws. This became a central aspect of our solution. We didn't just want to build a secure app; we wanted to create a tool that would allow women to maintain complete control over their health data and decide exactly how it was shared.

Through our partner's unique cloud technology, we developed an app that gave women control over their data in ways that weren't previously possible. Rather than storing sensitive data in third-party cloud servers, which could be vulnerable to breaches or misuse, our app ensured that the data was stored directly on the user's device. The user could then make active decisions about whether they wanted to share that data and with whom. By engaging in these once-uncomfortable conversations, we created a solution that empowered users with transparency and control, giving them the tools to protect their personal information.

The turning point in this project came when we stopped seeing these topics as difficult to talk about and started treating them as central to the conversation. By embracing the discomfort, we were able to design a tool that wasn't just technologically innovative but also deeply responsive to the needs and vulnerabilities of its users. And in the process, we shattered the taboos that had initially held our team back from fully engaging with the problem. This project was a powerful reminder that Equitable Design isn't just about addressing the technical aspects of a problem—it's also about having the courage to confront the sensitive, sometimes uncomfortable realities that lie beneath.

Notes

[1]   Lucky Iron Life, n.d. [Online]. Available at: `https://luckyironlife.com/?srsltid=AfmBOooW9HrgdOse9sBImUbclmVNORrKvfpGCogqtCTlJ8O40d9fS3a0.`

[2]   Dalal, M., 2014. Lucky Iron Fish in cooking pots tackle anemia. CBC. [Online]. Available at: `https://www.cbc.ca/news/health/lucky-iron-fish-in-cooking-pots-tackle-anemia-1.2658632.`

[3]   Eames Office, 1977. Powers of ten. [Online video]. Available at: `https://www.youtube.com/watch?v=0fKBhvDjuy0.`

[4]   Obreschkow, D., 2018. Universe size comparison: cosmic eye. [Online video]. Available at: `https://www.youtube.com/watch?v=8Are9dDbW24.`

[5]   Christensen Institute, n.d. Jobs to Be Done Theory. [Online]. Available at: `https://www.christenseninstitute.org/theory/jobs-to-be-done.`

[6]   Metinko, C. & Teare, G., 2024. Drop in venture funding to black-founded startups greatly outpaces market decline. Crunchbase News. [Online]. Available at: `https://news.crunchbase.com/diversity/venture-funding-black-founded-startups-2023- data.`

[7]   The Black Report, n.d. Funding. [Online]. Available at: `https://theblack.report/funding.`

Inclusive Design in Digital and Physical Spaces

Inclusion in design goes far beyond the technical details of a product—it touches on the human experience as a whole. Inclusive Design is about creating environments, systems, and tools that feel welcoming and usable for everyone, regardless of their background, ability, or circumstances. This chapter explores the nuanced layers of Inclusive Design, demonstrating how thoughtful implementation can enhance physical and digital interactions, spark systemic change, and redefine standards for equity and accessibility.

Whether through innovative banking experiences, reimagined packaging, or equitable onboarding processes, the stories in this chapter illustrate how inclusive solutions can have a profound impact on both users and businesses. These examples highlight the transformative power of addressing every touchpoint in the user journey. By doing so, companies not only unlock new opportunities but also ensure their solutions genuinely foster belonging and empowerment.

Through the lens of real-world examples and actionable strategies, this chapter invites you to think beyond the surface of design. It challenges conventional boundaries, illustrating how equitable solutions—when applied holistically—can break barriers, create connections, and lead industries toward more inclusive futures.

Think Beyond the Product, Look at the Experience

When it comes to designing for equity and inclusion, it's not just about the product itself but rather about the entire experience surrounding it. Every interaction a user has—from the packaging they see, to the way they navigate a service, to the support they receive—shapes their overall perception and determines whether they feel welcomed or excluded. Often, designers focus on making individual features more accessible, but it's the cumulative effect of each touchpoint that truly defines an inclusive experience. By examining and reimagining every step of the user journey, we can uncover hidden barriers and create environments where everyone feels valued and empowered. This section explores how looking beyond the product and considering the entire experience can lead to more meaningful and inclusive designs.

A strong example of this principle in action is the Guinness Storehouse in Dublin, a major cultural and tourist attraction that took deliberate steps to create a more inclusive visitor experience for neurodivergent individuals. Partnering with Ireland's national autism charity, AsIAm, the museum introduced Sensory Friendly Hours (see Figure 5.1)—specific time slots designed to provide a calmer and more predictable environment. During these hours, noise levels are reduced, lighting is adjusted, and trained staff are available to assist visitors. In addition, the Guinness Storehouse provides sensory kits—which include earplugs and visual guides—to help visitors feel more comfortable navigating the space. A sensory map of the museum allows individuals to plan their visit in advance, reducing anxiety and uncertainty.

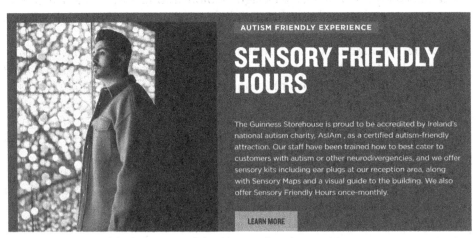

Figure 5.1: Extract of the homepage of the Guinness Storehouse visit center

This initiative exemplifies how businesses can go beyond compliance and truly design for comfort, autonomy, and dignity in public spaces. By acknowledging that neurodivergent individuals experience environments differently and making simple yet meaningful adjustments, the Guinness Storehouse creates a welcoming atmosphere for all guests—not just those with sensory sensitivities. In doing so, it enhances its reputation as an inclusive, forward-thinking institution while also expanding its potential audience.

Designing for True Inclusion: Addressing Every Touchpoint to Build Equitable Experiences

When I introduce the concept of Equitable Design, I often find that people have a hard time understanding how deeply it can shape products, services, and processes in ways that are truly inclusive. Most people's minds go straight to ergonomic solutions for individuals with physical disabilities—like jar openers for older adults or wheelchair ramps in public spaces. Although these examples are certainly important, they represent only a small part of what Equitable Design can achieve.

One example I always return to when explaining Equitable Design is the banking industry. On the surface, U.S. banks don't appear to be discriminatory. Yet if you dig into the numbers, a glaring issue emerges: 36% of African Americans are either underbanked or unbanked.[1] This raises a critical question: how is it possible that such a large portion of the population is excluded from accessing banking services, especially when these institutions claim to serve everyone?

Some time ago, I worked with a large bank to explore this issue in more depth. During our research phase, we conducted extensive interviews with both users and nonusers of banking services to uncover the barriers they faced. We didn't stop at talking to customers—we went out into the field to observe the experience of opening a bank account from start to finish.

What we discovered was eye-opening. For instance, we visited a local bank branch in a rural community, where we were met with the first major hurdle: inaccessible operating hours. The sign at the entrance read:

- **Monday: 10:00am – 11:30am**
- **Tuesday: Closed**
- **Wednesday: 10:00am – 11:30am**
- **Thursday: 10:00am – 11:30am**
- **Friday: 10:00am – 11:30am**
- **Saturday: 9:30am – 11:30am**

For anyone working regular business hours, these limited windows made it nearly impossible to visit the bank. Even for those who managed to get there, the atmosphere inside was far from welcoming. The entrance was guarded by a security officer, and the layout was confusing: long lines at the cashier on one side and a maze of cubicles on the other. It was clear that the process was designed more for the convenience of the bank than for the customer, particularly first-time users.

When we finally sat down with a bank officer to go through the account-opening process, the experience became even more daunting. The paperwork was overwhelming—long contracts, fine print, and complex jargon that would be difficult for anyone unfamiliar with financial terms to comprehend. The banker, pressed for time, added subtle pressure to sign the documents quickly, further intensifying the stress of an already confusing situation.

Once the account was open, the journey didn't get any easier. Many services required customers to navigate online banking platforms, which assumed access to a smartphone or computer—something not all customers had. The security protocols for logging in were complex, and without prior experience or trust in the system, it was easy for people to feel completely excluded.

The final challenge came in the form of communications—statements, bills, and promotional offers that flooded mailboxes with financial jargon. Terms like *debit*, *credit*, *interest rates*, and *credit scores* were used without any explanation, making it nearly impossible for people with low financial literacy to fully understand the services they were paying for. This was particularly troubling for underbanked communities, where financial literacy tends to be lower, compounding the issue of exclusion.

The implications of this exclusion go far beyond just opening a bank account. Without access to loans, many people from these communities are unable to finance higher education, buy homes, or start businesses—three key pillars for economic mobility. And although the banks themselves may not be overtly racist, their products, services, and customer experiences unintentionally create systemic barriers that perpetuate inequality.

This is where Equitable Design comes in. Through our partnership with the bank, we took a hard look at each of these touchpoints—opening hours, customer service, paperwork, online access, and communications—and redesigned the entire experience. The goal was to create an environment where all users, regardless of their background, could feel included and empowered to engage with the bank confidently.

We reimagined the layout of the branches to make them more welcoming and easier to navigate. We simplified the account-opening process by reducing jargon in the contracts and offering more personalized assistance. We also introduced alternatives for people without access to smartphones, ensuring that no one was left behind in the shift to digital banking. By redesigning each of these touchpoints, we were able to build trust with underbanked communities and provide them with the resources they need to access the financial services that are so critical to their economic growth.

Ultimately, this project was a clear demonstration of how Equitable Design can uncover and dismantle hidden barriers in everyday services. Through thoughtful redesign, we were able to create a more inclusive banking experience that allowed communities, especially those traditionally left behind, to thrive.

This is a critical lesson in how we can design for greater inclusion and diversity, whether in the physical or digital world. Every touchpoint matters. If a single aspect of the experience is inaccessible or exclusionary, it can undermine all other efforts toward equity and inclusion.

Take, for instance, the Champagne company we visited earlier in this book. As we toured the site, we noticed several ramps designed for wheelchair access. They were constructed with the correct gradient and width, and at first glance, they appeared to serve their purpose well. However, one of the ramps led to a door, and beyond that door were stairs—just stairs. This design flaw rendered the otherwise-accessible two-mile path across the estate unusable for individuals with disabilities. It was a stark reminder that a single inaccessible element can break the entire chain of usability.

This isn't just a lesson in physical spaces—the same applies to digital experiences. Imagine an app that's beautifully designed, featuring intuitive navigation and accessibility features such as text-to-speech or large fonts. But if the payment gateway is complex or the terms and conditions are filled with legal jargon that's impossible for someone with low literacy to decipher, it creates a barrier. That one inaccessible touchpoint can alienate users and nullify the effort put into creating an inclusive experience.

The Champagne company's inaccessible ramp and the hypothetical app share the same problem: a broken chain in the experience. When we think about inclusivity in design, it's not enough to make 90% of an experience equitable. That final 10%, if left exclusive, can invalidate all of the effort and resources spent on the other parts. Just as one broken link weakens an entire chain, one overlooked touchpoint in a customer journey can prevent users from fully engaging with or benefiting from the service.

When we worked with the banking sector, the same principle applied. Leaving one aspect of the banking journey—whether in-person interactions, contract clarity, or digital navigation—exclusive or inaccessible would compromise the potential benefits for underbanked communities. We had to ensure that every touchpoint, from opening hours to paperwork design, was reconsidered and optimized for inclusion. This kind of comprehensive approach to Equitable Design is essential to create meaningful, lasting impact.

True Equitable Design demands a holistic view. It's about ensuring that every individual, regardless of their abilities, background, or financial literacy, can navigate the world—both physical and digital—without unnecessary barriers. Anything less would be like building a bridge that leads to nowhere.

Rethinking Shampoo: How Sensory Design Can Break Barriers and Promote Inclusivity

In a project I've mentioned before, we worked with a personal care brand to rethink its shampoo line and make it more inclusive and equitable. Our first task was to examine the barriers that something as simple as shampoo could create, and we quickly realized that the product was not as universal as it seemed. Shampoos are almost always marketed in gender-specific or age-specific ways—targeting men or women, or kids versus adults. We initially focused on the packaging and marketing, but as we dug deeper, we began to see that every element of the product reinforced narrow social expectations and stereotypes, from the scent to the texture and color of the shampoo itself.

Take the scent of the shampoo, for example. Shampoos for men typically feature musky, woody aromas, whereas women's shampoos are often floral or fruity. These scent choices are not neutral; they reflect and reinforce gender norms and can make users feel boxed into a certain category. For someone who doesn't identify strictly as male or female or who doesn't resonate with these traditional scent profiles, choosing a shampoo can feel like choosing a gender identity rather than a personal care product.

Then there's the texture and color of the shampoo. For women, the product tends to be a thick, silky liquid, often white or pastel-colored, communicating softness and femininity. Men's shampoos, on the other hand, are usually transparent or dark-colored and more foamy or bubbly, reinforcing a sense of ruggedness or practicality. Even though texture and color may seem like small, inconsequential details, they play a role in the user experience and, consciously or not, convey messages about whom the product is for.

We realized that if we only changed the packaging and branding, we wouldn't be addressing the full picture. The product itself—the scent, texture, and color—was communicating a message that would likely lead to rejection by potential users who didn't feel represented. For example, a transgender individual or someone who prefers gender-neutral products might find themselves uncomfortable with either product option, feeling alienated by the gendered signals sent through something as simple as the smell or feel of the shampoo.

Beyond gender, there's the matter of sensory inclusivity. Not everyone experiences scent, color, and texture the same way. Someone with a sensitivity to strong fragrances might avoid the product altogether if the scent is overpowering, whereas someone with visual impairments may rely heavily on texture to differentiate products. These aspects are critical for designing a more inclusive product experience. We couldn't ignore the tactile sensation of the product as well—how the shampoo feels in your hand and on your hair communicates a lot about its intended audience. Is it soft and luxurious, or quick-foaming and efficient? These physical qualities can make people feel like a product "belongs" to them or not.

In the end, it became clear that if we wanted to make this shampoo line truly inclusive, we had to go beyond surface-level changes. The entire sensory experience—from the way the product smelled to how it felt in the hand—had to be rethought. We needed to design for the broad spectrum of users, ensuring that no one felt excluded based on gender, age, sensory preferences, or abilities. This meant crafting products with neutral or customizable scents and creating formulations that didn't signal any particular gender or stereotype through their color or texture. Only by addressing every sense and aspect of the product could we truly create a more equitable and inclusive shampoo line that everyone felt welcome using.

In the world of personal care, small design choices can have a big impact. Whether it's scent, color, texture, or even the sound of the bottle cap clicking open, these sensory details shape how users interact with and relate to the product. By reconsidering each of these elements, we can break down barriers and create products that are truly for everyone.

Inclusive by Design: Starting from the Beginning

In addition to focusing on every touchpoint in the user experience, it's crucial to make inclusiveness a priority from the very beginning—not as an afterthought.

We once collaborated with one of the world's largest companies, well-known for its search engine, to make its phones more inclusive for people with cognitive disabilities. The first step in our process was straightforward: we tested one of the phones. We turned it on, logged in with an account, and navigated the main menu while examining the interface—paying close attention to details like the colors, font sizes, and button sizes. These factors seemed promising. Then a team member suggested that we check out the accessibility features to see how well the company had integrated support for those with cognitive challenges.

It took us a few moments to locate the feature—it wasn't far, just a few clicks away. But we had to pause and think, "Where would the developers have placed it? Is it in the main settings? Display settings? Is there a dedicated category for accessibility?" As we hovered over the accessibility option, we all stopped. One of us broke the silence, asking, "How would someone with a cognitive disability even get to this point without the accessibility features turned on?"

It felt like a Kafkaesque moment, akin to trying to set up internet service but needing an active internet connection to do so. We immediately realized that the most significant improvement the company could make wasn't in tweaking the interface's finer details—it was in changing the approach entirely.

The accessibility features needed to be available from the very beginning of the phone's setup process, not tucked away in a menu. They should be the default option or, at the very least, offered as one of the first steps in the onboarding process.

This insight highlights an important lesson in Inclusive and Equitable Design: can your users access your product spontaneously, or do they need help before they even begin? Are your design choices inherent in the user journey, or were they tacked on as an afterthought? By making inclusiveness a fundamental part of the experience from the start, you ensure that all users, regardless of their abilities, can engage with your product seamlessly.

In recent years, the rise of healthcare portals and mobile apps has revolutionized how patients manage their medical records, schedule appointments, and communicate with healthcare providers. These systems are incredibly convenient for tech-savvy users, but for many older adults, they can present a series of daunting challenges—starting with the onboarding process itself.

Imagine an older adult, Joan, who has just been told by her doctor that she can access her lab results and schedule follow-up appointments online using the clinic's patient portal. Joan is an intelligent and independent woman, but she isn't particularly comfortable with technology. The first step of logging in to the portal requires her to create an account, but the registration process is far from intuitive. She's asked to input a long and complex password, followed by several security questions, all before even seeing her medical information. Already feeling overwhelmed, Joan tries to continue but stumbles through each step, unsure if she's doing things correctly.

For people like Joan, this initial experience can be so off-putting that they may choose to give up before ever accessing the health services they need. What should have been a straightforward experience instead becomes a source of stress and frustration.

After creating the account, Joan is asked to verify her identity through a method she doesn't understand—perhaps an email confirmation or a two-factor authentication text message. Navigating between devices, especially if she's not comfortable with smartphones or email, can feel insurmountable. It's a classic example of technology assuming a certain level of digital literacy that simply isn't there for all users, particularly older adults who didn't grow up with the internet.

Even after completing these hurdles, the layout of the portal is often unfriendly—with small fonts, too much information packed onto one screen, and unclear instructions on how to proceed. Joan may eventually find her lab results, but understanding the medical terminology and navigating the system to ask questions or schedule follow-up appointments presents yet another layer of complexity. Without any visual cues, voice guidance, or in-app support, Joan may feel as though she's navigating a maze with no clear end in sight.

To address these kinds of challenges, healthcare portals should offer a much simpler onboarding process specifically tailored for older adults. This could include features such as guided tutorials with clear instructions and large icons, voice-guided navigation, and options to skip certain verification steps or delay them for later (perhaps allowing patients to complete them with help from clinic staff). Video tutorials or step-by-step visual walkthroughs could ease the anxiety of trying to learn something new while managing important health information.

Moreover, including an in-person assistance option—either at the clinic itself or through a dedicated helpline—would ensure that people like Joan have support when they encounter digital roadblocks. The key is to ensure that the first experience older adults have with the platform is simple and supportive so that they feel empowered to use the system independently.

Incorporating these Inclusive Design features would not only make healthcare more accessible for older adults but also encourage more frequent engagement with their health providers—resulting in better health outcomes overall. By making the onboarding process a priority, healthcare providers can ensure that no patient is left behind due to digital barriers.

This should obviously be done in a way that preserves the security of users and their data. Complex interfaces and processes are usually that complex because they serve a purpose and have a functional necessity; the challenge of Inclusive Design is to compose experiences that are both functionally effective and easy to use for all audiences. This tension is pretty standard across most of the interventions I did for corporate clients and always requires good communication with the other people involved in the development process.

This question becomes even more urgent when the product or service inherently targets people who are more likely to face challenges, just like this healthcare example, where older adults are the primary users, often more so than younger, healthier individuals. These older adults may have difficulties with technology, cognitive decline, or limited mobility, making it crucial to design healthcare portals or systems that accommodate their specific needs right from the start.

Similarly, immigration and visa services primarily cater to foreigners, many of whom are unfamiliar with the country's administrative processes and may struggle with the language. These systems are essential for integrating into a new country, but if the design doesn't account for these language and cultural barriers, it can prevent people from accessing critical services.

Another clear example is public-transportation ticketing machines in cities like Los Angeles, where many users are people who can't afford a car. These individuals often come from lower-income backgrounds with limited education. If the system is difficult to navigate or lacks multilingual support, it excludes the very people who rely on it most.

A similar situation exists in online financial aid systems for universities. Many students applying for financial assistance come from low-income or first-generation backgrounds, and navigating complicated forms filled with jargon and requirements can become a barrier to getting the help they need. The process itself becomes exclusionary if not designed with these users in mind, leading to missed opportunities for students who might otherwise thrive with the right support.

In all these cases, the question of Inclusive Design is not just important—it's vital. When the users of a system are disproportionately likely to face limitations, making accessibility and inclusiveness part of the default experience becomes even more essential.

Breaking Barriers: How Thoughtful Onboarding and Distribution Create More Inclusive Access

More and more startups are becoming deliberate and strategic about their onboarding processes, prioritizing simplicity and accessibility for all users. A key example of this is the rise of freemium models and referral programs in digital products. These approaches not only are beneficial for business growth but also reduce barriers to entry for specific user groups. Freemium options, along with special offers for students, low-income individuals, or unemployed users, help to bridge gaps in accessibility. From an inclusivity perspective, this allows users who may not have the financial means to access premium features to still engage with the product, making the offering more equitable.

For example, platforms like Spotify and Canva offer freemium versions that allow users to explore the product's core functionality for free. This reduces the risk for users who might otherwise be hesitant to pay up front for a service they aren't familiar with. By offering student discounts or free trials for unemployed individuals, these companies make their products more accessible to those facing financial constraints, giving them a chance to use services that could enhance their education, job prospects, or personal development.

In addition to freemium models, referral programs and community-based products have also proven effective in building trust and encouraging user adoption. When users endorse products through referral programs, they provide a layer of social proof that helps new users feel more confident in trying the service. People are far more likely to engage with a product recommended by a trusted source, especially when that source is a friend or family member. These promoters often act as informal ambassadors, providing personal demos and showing the real-world benefits of the product from a user's perspective. The peer-to-peer connection adds a sense of familiarity and shared experience, making the onboarding process smoother and less intimidating for new users.

For instance, products like Dropbox and Uber have grown through their referral programs, where existing users are incentivized to invite others, often sharing their personal experiences and acting as de facto product guides.

This strategy not only drives growth but also democratizes access to the product, especially when users from underserved communities are involved. A person from a similar background explaining the product in a familiar cultural context can be far more impactful than a standard onboarding tutorial.

Additionally, distribution channels can play a significant role in creating more equitable access if designed thoughtfully. When intermediaries or community partners are empowered to distribute products, they not only expand reach but also create job opportunities and support local economies. For example, Grameenphone, a telecommunications provider, worked with local entrepreneurs to sell mobile services in rural communities, giving those entrepreneurs a sustainable livelihood while reaching underserved populations. Similarly, microfinance platforms often rely on local community members to educate and onboard users, ensuring that the service is not only available but understood and trusted.

By considering equitable distribution channels, companies can ensure that their products reach marginalized or underserved populations more effectively, all while empowering local agents to act as ambassadors for the product. These strategies combine accessibility, trust-building, and community empowerment, creating a more inclusive ecosystem that benefits both the users and the intermediaries involved.

The Power of Interfaces: Connecting People Through Thoughtful Design

As I've continued to work on Inclusive and Equitable Design, one concept that has grown increasingly important to me is the notion of the interface. Although the word *interface* often brings to mind digital screens or user interfaces (UIs) for apps and websites, its meaning goes much deeper. An interface is essentially any point of interaction between a product, service, or system and its user. This can be physical, like the handle on a door, or digital, like the menus on your phone. In fact, every product and service acts as an interface—it's the connection between the designer or company and the end user.

Design is fundamentally about creating interfaces because it defines how users will engage with a product or service. An interface should be intuitive, easy to use, and reflective of the needs of the people interacting with it. The more thought and care put into an interface, the more likely it will resonate with users and deliver an experience that feels inclusive and considerate.

At its core, every product—whether it's an app, a physical item, or even a service—is an interface between the user and the designer or the company. Take, for example, a product like the iPhone. Most Apple users will never visit the company's headquarters in Cupertino or meet Tim Cook, but they feel connected to Apple through the company's products. The iPhone's design and interface communicate Apple's values, culture, and attention to detail. Steve Jobs famously emphasized that the level of detail and cleanliness in the design of the inside of a device should match the outside. His philosophy was that even if no one ever saw those inner components, they were a reflection of Apple's commitment to excellence. In this sense, the product itself becomes an interface that expresses the company's identity and its connection with its customers.

That's why it's critical for companies to view their products and services as more than just functional tools—they are connectors that allow companies to communicate their values, desires, and causes. When users interact with these products, they should feel as though they were considered and thought of during the design process. This leads to a sense of resonance, where users can connect with specific features or qualities that make them feel included or understood.

A powerful example of Inclusive Design can be seen in the project Le Reflet, a restaurant in Nantes, France, which promotes social and professional integration by employing both professionals and individuals with physical and cognitive disabilities. The goal of the restaurant is to create a space where people of all abilities can work together and share a unique culinary and human experience.

One of the most thoughtful design solutions implemented at Le Reflet involved the creation of specially designed plates (see Figures 5.2 and 5.3) to assist the restaurant's staff, many of whom have Down syndrome. Individuals with Down syndrome tend to have smaller hands and shorter fingers, which can affect their ability to grasp and stabilize objects. Recognizing this, the team designed plates with small cavities along the edges that offer a better grip for servers, helping them carry and handle the plates more securely.

However, these plates serve a dual purpose. Although designed to assist the staff, they also act as an important interface for the restaurant's customers. As guests interact with the plates, they are subtly drawn into the story of the restaurant's mission to promote inclusion. These thoughtfully crafted plates, with their ergonomic design, allow the servers to perform their duties confidently while also communicating a deeper narrative to the diners.

Figure 5.2: Plates with handprints designed by Pulse and Pulpe design agency for Le Reflet Restaurant in France

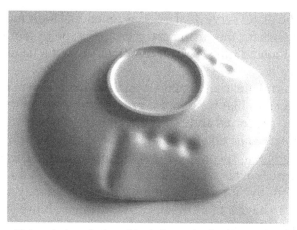

Figure 5.3: Plates with handprints designed by Pulse and Pulpe design agency for Le Reflet Restaurant in France

For the customers, the plates become more than just functional objects; they represent a tangible connection to the values of Le Reflet. The unique design, shaped by the needs of the servers, subtly conveys a message about accessibility, inclusion, and empathy. Diners not only experience the delicious food but also are invited to reflect on the inclusive environment they are part of, making the meal both a culinary and human experience.

The design process behind the plates was highly intentional. Several tests were conducted with individuals with Down syndrome to ensure that the final design would genuinely enhance their ability to hold the plates. These plates are not only functional but also aesthetically pleasing, blending accessibility with visual appeal. The goal was to create a design that supported the needs of the workers without highlighting their disabilities, ensuring that the plates fit seamlessly into the restaurant's environment.

The role of the design agency Pulse & Pulpe was instrumental in transforming the initial concept into a functional product. The agency worked on the technical 3D design of the plates based on the initial sketches and collaborated with manufacturers to prototype and produce the plates. This thoughtful and collaborative process shows how Inclusive Design can go beyond functionality to also embody beauty and dignity for those who use it. The result is a holistic design that meets the practical needs of the servers while adding an aesthetic element to the dining experience.

By placing these specially designed plates in front of the diners, the restaurant turns a simple object into a storytelling tool. Customers feel and see the intentionality behind the design, creating a meaningful interaction between the object and the user. The plates thus act as an interface—not just between the servers and the product but between the restaurant and its guests. Through this interface, Le Reflet's commitment to equity and dignity is clearly communicated, elevating the dining experience into a moment of shared understanding and connection.

In this way, the plates serve dual roles: as functional tools for the servers and as symbolic artifacts for the customers, reinforcing the restaurant's mission of fostering inclusion and offering everyone a place at the table—both literally and metaphorically.

Another powerful illustration of this concept can be seen in public transportation design. In cities like London, for example, the "mind the gap" markers between subway platforms and trains are an important safety feature. But in recent years, the city has taken further steps by implementing tactile paving along platform edges for people with visual impairments. This not only serves as a physical barrier but also as an interface that makes the subway system more accessible and intuitive for a wider range of users. What may seem like a small design change can have a huge impact on someone's ability to independently navigate their environment.

Ultimately, the goal of Inclusive Design is to ensure that users of all abilities and backgrounds can engage with products in a way that feels natural and considerate. By focusing on every detail of the interface, designers can make sure that users feel seen, valued, and empowered through their interactions. Interfaces are the language of design, and when crafted thoughtfully, they can bridge gaps between users and experiences, making the world a more equitable place for all.

Empowering Communities: How Companies Can Redefine Their Role in Society

In a world facing unprecedented social and environmental challenges, businesses have the potential to be powerful agents of change. Traditionally, companies have focused on minimizing their negative impacts through internal improvements, such as reducing waste or adopting sustainable practices. But what if they could go further? By leveraging their unique strengths, resources, and expertise, companies can transcend the boundaries of their operations and actively contribute to societal progress. This section explores how businesses can redefine their role in society, using their products, technologies, and networks not just to serve their bottom line but also to empower communities and drive systemic change. Through a series of impactful projects, we'll see how companies can become catalysts for a more equitable and inclusive world.

Putting Your Product's Strength at the Service of a Cause

A while back, I had an engaging conversation with a senior executive from one of the world's largest electronic device companies. Despite the company's overwhelming success and significant investments in R&D, our discussion centered around a pressing issue that is close to my heart: sustainability and the urgent need to reduce plastic pollution. I was eager to explore how such a powerful company could take tangible steps to decrease its plastic usage and consumption. However, the executive's response was both candid and thought-provoking.

He said, "Look, we can't eliminate plastic from our supply chain or products entirely. It would be misleading to claim that we're a 'nonplastic' company just because we've managed to reduce 0.2% of the plastic in our products. But we have incredible resources in R&D, and I believe we could make a much greater impact by leveraging our expertise and capabilities to serve the cause more effectively. Imagine if we developed technologies like robots, IoT solutions, or software systems specifically designed to aid in sorting, recollecting, reusing, and recycling plastic."

His words were a revelation, offering a refreshing perspective that has stayed with me ever since. It's not always about striving for unattainable purity by eliminating the use of a problematic material like plastic, especially when operational constraints are so tight. Instead, it's about finding innovative ways to apply a company's unique strengths and resources toward a greater cause. If a company cannot eliminate its harmful impact directly, it can still use its power and expertise to drive broader change.

This idea came up again during a conversation with a top executive at a major bank. The executive argued that the bank wasn't a significant contributor to plastic pollution, apart from small changes like eliminating plastic cups in office kitchens and adding recycling bins. Although these efforts were commendable, they felt almost symbolic compared to the bank's potential impact.

We began to explore a different approach: how could the bank's resources—specifically its financial power and network—be leveraged to combat pollution more effectively? Instead of focusing solely on reducing its direct footprint, we proposed that the bank channel a portion of its loans toward innovative start-ups and companies developing sustainable technologies. We also discussed the possibility of connecting the bank's clients with each other to foster collaborations focused on environmental solutions. The idea was to use the bank's vast influence not just for its own sustainability efforts but also to become a catalyst for broader change within the industries it served.

These conversations taught me that sometimes, tackling a massive issue like plastic pollution requires thinking beyond immediate self-correction. It's about recognizing the full scope of your capabilities and using them to drive systemic change. Whether it's a tech giant using its R&D to develop revolutionary recycling technology or a financial institution using its capital to support sustainable ventures, the key is to harness what you do best and apply it where it can make the biggest difference. This mindset shift has become a cornerstone of how I approach challenges in Equitable and Inclusive Design, always looking for ways to align strengths with meaningful impact.

Leveraging Strengths for Greater Impact: The Writing Assistant and Team Collaboration Platform Projects

In our work with a company renowned for its digital tool that helps users improve their writing, we encountered a unique opportunity to leverage the platform's strengths to address language discrimination and inclusivity. Instead of merely refining the existing product, we explored how this tool could utilize its core capabilities to make a broader societal impact.

The challenge was clear: language can be a powerful tool for inclusion, but it can also unintentionally perpetuate biases and discrimination. This is especially true in diverse teams where cultural nuances and linguistic differences can lead to misunderstandings and unintentional exclusion. Our goal was to transform this writing tool into a resource that not only corrects grammar but also promotes inclusive language in professional environments.

We proposed a solution that integrated a database of inclusive language guidelines, helping managers and teams identify potentially discriminatory language patterns. This approach went beyond traditional spelling and grammar checks, enabling the tool to serve as a guide for more respectful and inclusive communication within organizations. By utilizing its robust data capabilities, the tool could track language usage over time, providing insights and recommendations to foster a more inclusive and respectful culture.

This project exemplifies how a company can leverage its existing strengths—in this case, its expertise in language and technology—to address broader social issues. Rather than overhauling its own practices, the company used its platform to empower others to communicate more inclusively.

Similarly, in our collaboration with a leading provider of project management and collaboration software, known for its tools that help agile teams organize and track their work, we faced the challenge of optimizing its product to enhance inclusion and belonging within those teams. Although its tools are invaluable for managing projects and tasks, they can inadvertently foster competitiveness, conflict, and biases if not used thoughtfully.

Our solution involved creating a feature that allows managers to assess the sense of inclusion and belonging within their teams. By developing a set of key performance indicators (KPIs) that measure team dynamics and collaboration, the platform could provide managers with data-driven insights to make more informed, inclusive decisions. This project highlighted the importance of using existing strengths—in this case, data analytics and project management expertise—to foster a more inclusive workplace.

Both of these projects illustrate a key principle in Equitable Design: even when a company's core practices may not seem like the obvious tools for addressing social issues, those very strengths can be leveraged to create a significant impact. It's not always about changing what you do; sometimes, it's about reimagining how you can use your expertise to serve a greater purpose.

Finding Places to Share the Power to Uplift Underserved Communities

The previous example illustrates how an existing platform and product can be leveraged to promote more equitable practices and behaviors. However, we can take this concept even further by exploring how technology can be used to redistribute power across communities.

A core principle of Equitable Design is finding those moments where a product or service can go beyond its primary function to actively facilitate a more balanced distribution of power. It's about creating systems that not only address current inequalities but also empower diverse groups to have a stake in shaping the future.

The project we worked on offered a unique opportunity to explore the potential of Web3 technology in fostering a more equitable and inclusive economic system. The central idea was to leverage blockchain capabilities not just to enhance company practices but to fundamentally shift how value and power are distributed within communities.

Imagine a startup that wants to create a more inclusive business model but doesn't know how to engage its diverse community effectively. Typically, companies might focus on making internal changes, like improving diversity within their teams or adopting more inclusive language. However, the challenge with these approaches is that they are often limited to the organization itself, with a relatively small direct impact on the broader community.

This is where blockchain technology comes into play. It enables companies to go beyond internal changes and extend the benefits to a larger ecosystem. By implementing a decentralized system, companies can allow community members, users, and even customers to become shareholders. This creates a system where the community can have a say in the company's direction and share in its financial success.

One of the most exciting aspects of this project was the concept of Community Tokens—a groundbreaking way for companies to share their success with the very people who make it possible. These tokens could be distributed to users and community members based on their engagement, contributions, and advocacy for the brand, effectively turning customers and collaborators into stakeholders. Imagine how transformative this could be for car-sharing platforms, where the business model relies not just on the employees but also on a vast network of independent drivers and loyal riders.

Consider the independent drivers who make car-sharing services viable. They are the backbone of the operation, yet they typically don't see much beyond their immediate earnings. What if they could be rewarded with tokens that represent a stake in the company, giving them a share in its growth and success? Similarly, users and riders, who are essential to generating revenue for both the company and the drivers, could also be empowered. By offering them tokens, the company acknowledges their role in the ecosystem, incentivizing them to continue supporting and promoting the platform.

This approach goes beyond traditional incentives and loyalty programs; it's about shifting from a shareholder economy to a stakeholder economy. It recognizes that the success of a business depends on a broader community of contributors—drivers, users, advocates—and aims to reward them not just as customers but as partners in growth. By creating this inclusive framework, we enable companies to build deeper connections with their communities, fostering loyalty and shared prosperity.

The impact of such a system is profound. It enables companies to share their financial upside with those who help create it, thus aligning incentives in a way that traditional business models rarely do. Moreover, it gives underrepresented communities a tangible stake in the success of the businesses they support. This can be especially transformative for communities that have historically been excluded from traditional financial systems.

This example illustrates how blockchain technology can be more than just a tool for financial transactions; it can be a powerful mechanism for social impact. By reimagining how value and power are distributed, we can create more equitable systems that go beyond the limitations of traditional business practices. This platform provided the perfect vehicle to explore this potential, enabling companies to use their technological strengths to improve their practices as well as create lasting, positive change in the communities they serve.

Introducing Technology at the Same Pace to Avoid Deepening the Gap

A direct approach to contributing to a more equitable society through design is to ensure that solutions are made accessible to everyone simultaneously, regardless of socioeconomic status or geographic location. Often, inequalities arise because new technologies or services are rolled out unevenly, with wealthier neighborhoods and larger organizations being the first to receive the best resources. This early access to superior technology and services further cements their advantageous position, providing them with better opportunities and efficiencies, while underserved communities are left to wait years before they can access comparable resources. This delay places an additional burden on these communities, making it even harder for them to compete and thrive in an open market.

A stark example of this disparity can be seen in San Francisco, a city often perceived as the global tech capital. Despite its reputation, there are significant inequalities in internet access across the city. According to the San Francisco Tech Council, 91.6% of San Francisco households have a broadband internet connection, but connectivity across the city is not uniform. There are many areas where internet connections are unreliable or unavailable, especially in older buildings and public housing units, where connectivity issues disproportionately impact low-income families, seniors, and people with disabilities.[2]

This issue is particularly evident in neighborhoods like Chinatown, Bayview Hunters Point, Sunnydale, and Potrero Hill, where residents struggle with connectivity issues ranging from dead zones to unreliable internet service. For example, in Chinatown, only 63% of residents have a broadband plan of any kind, with many businesses still relying on outdated dial-up connections. The barriers are even more pronounced in the approximately 530 single-room occupancy hotels (SROs) in San Francisco, housing about 22,000 people, many of whom lack internet service in their buildings and cannot afford high-cost subscriptions.

The digital divide is not just about availability but also about adoption. Although 35% of San Francisco households are eligible for the Affordable Connectivity Program (ACP), only 37% of those households are enrolled, far below the statewide enrollment rate of 49%. The complex enrollment process and lack of awareness are significant barriers, particularly for older adults and people with disabilities. These populations are often unaware of the benefits they are eligible for, and the digital divide is further exacerbated by language barriers, as nearly half of adults aged 60 and older in San Francisco speak a primary language other than English.

This situation underscores the importance of a more equitable approach to deploying technology. Companies and policymakers should prioritize ensuring that all communities have access to reliable internet and the skills and tools needed to navigate the digital world. By investing in enhanced internet infrastructure, digital literacy training, and accessible technology, we can help bridge the digital divide. The SF Tech Council's Digital Equity Plan for Older Adults & Adults with Disabilities (2024–2028) lays out a comprehensive strategy to address these challenges, including recommendations to expand free and affordable internet options, develop a robust device ecosystem, and integrate digital literacy and skills training into workforce development programs.

This highlights the need for a more equitable approach to deploying technology. When developing new products or services, companies should consider how they can be made accessible to all segments of society from the outset. Whether it's ensuring that internet access is treated as a utility or designing platforms that are inclusive of all user groups, these considerations are crucial for creating a more just and equitable society. By prioritizing equal access and support for underserved communities, we can help bridge the digital divide and enable more people to participate in and benefit from the opportunities of the digital age.

Equitable Deployment of Technology: The Electric School Bus Project

In our collaboration with a global energy leader, we faced a unique challenge: how to make clean, electric transportation accessible to all school districts in California at the same pace. This wasn't just about providing a more environmentally friendly option but also about ensuring that every community, regardless of its economic standing, could benefit from the health and environmental advantages of electric buses.

The goal of this project was to bridge the gap between this energy company and California's school districts, enabling them to transition from traditional diesel-powered buses to electric ones. We focused on understanding the specific barriers that different schools faced when adopting this technology. Wealthier districts, typically located in more affluent areas, were often quicker to adopt electric buses, benefiting from reduced emissions and lower operating costs. However, many lower-income school districts, where pollution levels are often higher and funding is more limited, were left behind in this transition. This disparity not only perpetuated environmental inequities but also worsened health outcomes for students in these underserved communities.

To address this, we developed a comprehensive strategy that included gathering detailed data on existing school bus fleets, the demographics of the school districts, and the health impacts of diesel emissions on students. We created a prototype assessment report that provided personalized recommendations for each school district. This report included information on fleet age, efficiency, time, energy consumption, and the potential health benefits of transitioning to electric buses. The idea was to create a compelling, data-driven case for the adoption of electric buses that was tailored to each district's specific needs and constraints.

One of the most impactful insights came from our research into the transportation patterns of low-income families. For many parents in these communities, the school bus is not just a convenience but a necessity. They rely on it as their primary mode of transportation for their children because they cannot afford alternatives like driving them to school. However, the older diesel buses used in these districts contributed significantly to air pollution, posing serious health risks to the children who used them daily.

To ensure an equitable deployment of technology, we knew we had to address more than just the cost of the buses. We needed to engage a wide range of stakeholders, including school administrators, parents, transportation coordinators, and community leaders, to create a sense of shared ownership and urgency. We also looked at innovative financing options, such as leveraging grants and government incentives, to make it easier for lower-income districts to adopt electric buses without straining their already tight budgets.

By focusing on an inclusive and comprehensive approach, we aimed to create a model where every school district, regardless of its financial standing, could transition to cleaner transportation at the same pace. This project highlights the importance not just of introducing new technologies but also of ensuring that they are deployed equitably so that all communities can benefit from the advantages they offer. It's about more than technology; it's about creating systems and processes that actively work to close the gaps and reduce inequalities.

Bridging the Digital Divide for Small Businesses: The Virtual Assistant Project

In collaboration with a global technology leader, we set out to tackle a critical issue facing small businesses: the digital divide. Although large corporations often have the resources and expertise to integrate advanced technologies into their operations, small businesses are frequently left behind, struggling to keep up in an increasingly digital world. This project aimed to leverage advanced virtual assistant tools to empower small businesses, helping them not only to survive but thrive in today's competitive landscape.

Meet Maria, a small business owner in California. Like many of her peers, she faces significant challenges in adopting and effectively utilizing digital technologies. According to our research, 31% of small business owners in America try to use technology but aren't sure how to use it most effectively, and 42% don't use technology to its fullest capacity. For Maria, this gap translates into lost opportunities, inefficiencies, and a constant struggle to keep up with the digital demands of her customers and the marketplace.

To address these challenges, we developed a solution centered around a virtual assistant tool specifically tailored to the needs of small businesses. This tool, referred to as AI for SME, was designed to integrate seamlessly into business websites, providing personalized and relevant prompts to help business owners like Maria navigate the complexities of digital integration. The goal was to make digital adoption not just easier but also intuitive and aligned with the unique needs of each business.

The project was guided by three main objectives: customization, trust, and revenue generation. Customization was essential to ensure that the digital solutions offered were tailored to the specific needs of each business. By building trust in technology, we aimed to overcome the skepticism and hesitation that many small business owners feel toward adopting new digital tools. And by enhancing the shopping experience for customers, we sought to help these businesses generate additional revenue.

AI for SME offered several key features to achieve these goals. It provided a simple, AI-powered search engine that could extract insights from unstructured data, making it easier for business owners to find and utilize relevant information.

It also facilitated learning through conversations and instructions, requiring no technical knowledge on the part of the user. Importantly, the tool was designed with a human-centric approach, ensuring data protection and using a genuine, authentic tone in all interactions. This focus on user-friendliness and privacy was crucial for building trust among small business owners.

One compelling use case came from a family-owned restaurant. The owner, who previously had to manage online orders and reservation requests manually, was able to use AI for SME to automate these processes. This not only freed up time for the owner to focus on other aspects of the business but also significantly improved the customer experience by providing quicker and more reliable service. For small business owners who are often overwhelmed by the demands of running their operations, this kind of support can be transformative.

Our approach was to ensure that small businesses could adopt this technology at the same pace as larger enterprises, thereby reducing the digital gap that often puts them at a disadvantage. This project wasn't just about introducing new tools; it was about making sure those tools were accessible, affordable, and truly beneficial for small businesses.

Transforming Technology: The Essential Role of Hardware and Software in Inclusive Design

As an entrepreneur in Silicon Valley, I've come to appreciate the unique advantages that software offers when it comes to building a startup. For many industrial designers, creating innovative hardware products remains the ultimate goal—a tangible manifestation of their skills and creativity. But for engineers and business leaders, the dream of developing a software company has often held a special allure. Software represents the promise of rapid innovation, scalability, and the ability to reach millions of users with minimal overhead. It's a realm where a small team with a great idea can disrupt entire industries from a single office.

This contrast between hardware and software in the startup world is stark. Hardware development involves a complex and costly process: prototyping, sourcing materials, manufacturing, logistics, and distribution. Each iteration requires significant investment and time, and any error in design or functionality can be financially devastating. Moreover, once a hardware product is released, making changes is incredibly difficult. This rigidity stands in sharp contrast to the fluid nature of software development.

The beauty of software lies in its flexibility and cost-effectiveness. It's relatively inexpensive to develop, particularly in the initial stages. With the right team and tools, entrepreneurs can build and launch a product from virtually anywhere, iterating and refining it based on user feedback. Software can be updated, patched, and improved in real time, enabling a level of personalization and responsiveness that hardware struggles to match. This dynamic nature means that software products can evolve with their users, adapting to their needs and preferences almost instantly.

For these reasons, making digital tools and experiences more inclusive should be a straightforward endeavor. Software can be designed to accommodate a wide range of users from the outset, incorporating features like adjustable text sizes, voice commands, and customizable interfaces to cater to different needs and abilities. Updates can be rolled out to enhance accessibility, ensuring that no user is left behind as the product evolves. In contrast, making hardware more inclusive often requires significant redesign and reengineering, which can be prohibitively expensive and time-consuming.

Our project with a leading technology company focused on a significant challenge: how to make one of the most widely used office productivity suites more accessible and inclusive for neurodiverse individuals. The goal was not only to improve usability but also to address deeper issues, such as the stigma around requesting accommodations and the internalization of struggles that many neurodiverse individuals face in professional environments.

The challenge we encountered was twofold. On the one hand, there was a need to redesign the user interface of the office suite to better cater to the unique cognitive needs of neurodiverse users. On the other hand, we had to shift the perception of accessibility from being an afterthought or a secondary feature to becoming an integral part of the product's core design. This meant rethinking how features were presented and accessed, ensuring that the software was intuitive and comfortable for all users.

Objective: Designing for Cognitive Inclusion

The objective was clear: reimagine the software suite to make it more accessible for neurodiverse individuals. This meant going beyond traditional accessibility features and creating a workspace where every user could thrive in their own way. Our approach focused on simplifying the interface, reducing cognitive load, and providing personalized customization options.

One of the key features introduced was a minimized ribbon view. The traditional ribbon interface, although powerful, can be overwhelming due to the sheer number of options presented simultaneously. By offering a minimized version, we allowed users to focus on the essential tools they need without being distracted or confused by unnecessary tabs and options.

Additionally, we introduced grouped functionalities, which combined commonly used tools into convenient clusters. This not only made the interface cleaner and less daunting but also reduced the number of steps required to perform routine tasks, making the software more intuitive and less frustrating for users who might struggle with navigating complex menus.

Recognizing that no two users are the same, we also incorporated deeper customization options. Neurodiverse individuals often have unique preferences and needs when it comes to how information is presented. Some may prefer high-contrast themes, whereas others may need larger text or specific color schemes to avoid sensory overload. The redesigned interface allowed users to craft their own personalized workspace or even enlist AI assistance to guide them in building a setup that best supported their workflow.

This level of customization extended beyond aesthetic preferences. We introduced features that could adapt to the user's working style, such as adjustable task lists, simplified toolbars, and customizable shortcuts. These tools empowered users to create a workspace that not only accommodated their needs but also enhanced their productivity and comfort.

One of the most impactful aspects of this project was establishing a continuous feedback loop. We engaged directly with neurodiverse individuals throughout the design process, incorporating their input to refine and improve the software. This approach helped us move beyond assumptions and ensure that the final product truly resonated with the community it was designed to serve.

The significance of this project extends beyond software development. It represents a shift toward a more inclusive digital landscape, where technology adapts to the needs of its users rather than expecting users to conform to the limitations of the technology. It highlights the importance of making digital tools and experiences more inclusive and underscores that inclusivity should be a fundamental consideration from the outset.

This project exemplifies how software, with its inherent flexibility and adaptability, can lead the way in creating more equitable and accessible experiences. By leveraging the power of technology to meet the diverse needs of all users, we can foster an environment where everyone, regardless of their cognitive abilities, can contribute and succeed. This is the essence of Equitable Design—building products that empower individuals and create opportunities for all.

Bridging the Digital Divide: Integrating Hardware and Software for Inclusive Innovation

Although software has dominated the startup landscape in Silicon Valley due to its flexibility and cost-effectiveness, this focus has sometimes narrowed our view of what true innovation can be. The ease of developing and deploying software solutions often overshadows the significant potential of hardware products that integrate seamlessly with digital technologies. Some of the most impactful innovations of our time—smartphones, wearable devices, and electric vehicles—have demonstrated that when hardware and software work together, they can create profound and life-enhancing experiences. This project with a leading technology company aimed to highlight exactly that potential by reimagining a smartwatch to be more inclusive for people with visual impairments.

The project was centered around making this smartwatch not just a piece of technology but a meaningful and empowering tool for those with visual impairments. The goal was to enhance the day-to-day functionality of the smartwatch by ensuring that every element—from setup to daily use—was designed with accessibility in mind. This required a holistic approach, looking at both the software and the hardware aspects of the device and how they interact within the larger ecosystem of connected devices.

The Challenge: Accessibility Beyond the Screen

The primary challenge was to make the smartwatch accessible and inclusive in a way that would allow users with visual impairments to navigate and utilize its features independently and confidently. This required more than just software adjustments like text-to-speech functionalities; it meant rethinking the physical design and how the device communicated with the user.

For many visually impaired users, technology can often feel like an additional hurdle rather than a helpful tool. Complex interfaces, tiny buttons, and the reliance on visual cues make many devices difficult to use. The project team recognized that to truly empower these users, they needed to create a device that was intuitive and accessible from the moment it was unboxed.

The Solution: Integrating Hardware and Software for Inclusive Design

The team implemented several key solutions that integrated hardware and software seamlessly. For setup and unboxing, they included braille on packaging and instructions, as well as auditory cues during the initial setup process, making it possible for visually impaired users to start using their devices independently from the very beginning. This small but significant change demonstrated that inclusivity begins not when the product is turned on but the moment it is taken out of the box.

The physical design of the smartwatch was also enhanced with snap-on band technology, making it easier to attach and more stable on the wrist. This small hardware modification significantly improved usability for those who might struggle with traditional watch straps.

On the software side, the team simplified the user interface, reducing the number of button presses required to access different functionalities. This was coupled with enhanced connectivity to the company's broader ecosystem of devices, using technologies like Bluetooth proximity, Wi-Fi networks, and Ultra Wideband sensing to enable features such as automatic recognition of nearby devices and voice-activated commands. For example, the smartwatch could connect with a smart refrigerator or oven, providing the user with tactile or auditory feedback about the status of these appliances. This extended the smartwatch's functionality beyond a standalone device, integrating it into a supportive, connected ecosystem that could adapt to the user's needs.

One particularly innovative feature was the implementation of braille haptics. By using a combination of vibrations and haptic feedback, the smartwatch could communicate information such as time, notifications, and navigation prompts in a tactile language that the user could feel. This created a unique, nonvisual interface that allowed users to access critical information discreetly and effectively.

The Importance of Ecosystems in Inclusive Innovation

This project is a powerful reminder that true innovation requires looking beyond isolated products to consider the entire ecosystem of devices and technologies that interact with one another. The smartwatch was not just an isolated piece of hardware but a part of a broader network of connected devices that could work together to create a more inclusive environment for the user. By leveraging the existing ecosystem of connected appliances, home automation systems, and mobile devices, the team was able to create a solution that was greater than the sum of its parts.

This project exemplifies the importance of integrating hardware and software in the development of inclusive products. Although software solutions like voice commands and screen readers are essential, they are most powerful when combined with thoughtfully designed hardware that considers the physical interactions users have with their devices. In this case, the smartwatch became more than just a digital tool; it became a bridge between the user and their environment, providing a tangible, intuitive interface that made everyday tasks more accessible.

The success of this project demonstrates that innovation in hardware is not only possible but necessary in creating truly inclusive technologies. By designing products that leverage the strengths of both hardware and software, we can build a future where technology is accessible to everyone, regardless of their abilities. This holistic approach to design ensures that we are not just creating devices but empowering tools that enrich the lives of all users.

Notes

[1] Federal Deposit Insurance Corporation (FDIC), 2021. Appendix to the 2021 FDIC National Survey of Unbanked and Underbanked Households. [Online PDF]. Available at: `https://www.fdic.gov/analysis/household-survey/2021appendix.pdf` [Accessed: 16 January 2025].

[2] SF Tech Council, 2023. SF Digital Equity Plan for Older Adults & Adults with Disabilities. [Online]. Available at: `https://www.sftechcouncil.org` [Accessed: 16 January 2025].

Designing for, with, and by Communities

Designing for communities—and, more importantly, with them—requires more than just empathy or good intentions. It demands thoughtful engagement, careful consideration of power dynamics, and the integration of established methodologies tailored to foster equity. Although many of the strategies explored in this chapter, such as rapid prototyping, contextual interviewing, and photo-elicitation, are standard tools in the design world, they take on new significance when applied through an equitable lens.

The practices discussed here extend beyond their conventional applications by focusing on empowerment and inclusion, ensuring that marginalized voices are not only heard but also actively shape the outcomes. This approach benefits the communities in question and leads to innovations that have universal appeal and utility, highlighting a core thesis of this book: Equitable Design serves everyone, not just a select few.

In this chapter, we'll explore three levels of community engagement—designing *for*, *with*, and *by* communities—each offering unique insights into navigating the complexities of Inclusive Design. You'll learn how to adapt established methods to the specific goals of Equitable Design, discover ways to bridge gaps between designers and communities and see how empowering users can lead to solutions that are not just functional but transformative. Through case studies, strategies, and lessons learned, we'll demonstrate how meaningful collaboration creates designs that are as impactful as they are inclusive.

Community Engagement in All Degrees

A crucial element of successful Inclusive and Equitable Design is meaningful engagement with the communities for whom the design is intended.

Before diving into examples and case studies, it's essential to unpack what "meaningful engagement" truly means and why it is a cornerstone of designing for and with communities. At its core, meaningful engagement is actively involving people—especially those directly impacted by a project—in the design process. It goes beyond consultation or surface-level input, focusing on building genuine, reciprocal relationships where every voice is heard and valued.

This approach is vital for several reasons. First, it ensures that solutions are grounded in the lived experiences of the people they are meant to serve. By drawing on the unique insights and expertise of community members, designers can uncover needs, challenges, and opportunities that might otherwise remain invisible. Second, meaningful engagement fosters trust and buy-in, increasing the likelihood that the resulting solutions will be embraced and sustained over time. Finally, it challenges traditional, top-down approaches to problem-solving by empowering communities to shape the outcomes that affect their lives.

Meaningful engagement is not without its challenges. It requires designers to listen deeply, approach with humility, and navigate complex dynamics of power and privilege. However, the benefits far outweigh the difficulties, as this process leads to solutions that are not only more equitable and effective but also more innovative. By starting with a clear understanding of this foundational concept, we can better appreciate the significance of the stories and strategies explored in this chapter.

As discussed earlier, designers often face challenges in truly understanding their target users because individuals from underserved communities, people with disabilities, and minorities are significantly underrepresented in design, creation, and product development roles.

This disconnect has led to numerous examples of projects failing due to a lack of alignment with the needs and realities of their intended audience.

One stark example of this disconnect is the PlayPump project, an initiative introduced in South Africa with the ambitious goal of addressing water scarcity in underserved communities. The concept was seemingly ingenious: a merry-go-round that, when spun by playing children, would pump water from underground into a storage tank, providing both a source of clean water and a playful environment for kids. The project received widespread acclaim and financial backing due to its innovative approach to a critical issue (see Figure 6.1).

The PlayPump was designed by Ronnie Stuiver, a borehole driller and engineer from South Africa, who first exhibited the concept at an agricultural fair in 1989. The commercialization of the PlayPump was led by Trevor Field, who licensed the idea and founded Roundabout Outdoor to manufacture and install the systems; the organization subsequently won the World Bank Development Marketplace Award in 2000. The initiative gained significant funding and support from organizations such as the Case Foundation, founded by AOL's Steve Case, which helped establish PlayPumps International (PPI-US) as a nonprofit organization to promote the technology in 2004.[1]

Figure 6.1: The PlayPump project in Africa

However, despite its initial success and media attention, the PlayPump quickly became a cautionary tale of how even the most well-intentioned designs can fail without genuine community engagement and understanding. The primary flaw lay in the assumption that children would be able and willing to play on the merry-go-round long enough to pump the required amount of water for the community. In reality, the amount of effort needed to generate sufficient water far exceeded what children could provide. This led to situations where adults had to use the equipment, an experience that was both impractical and undignified.

Moreover, the PlayPump project was developed and implemented with minimal input from the communities it aimed to serve. Local residents were not involved in the design process, and their unique needs and preferences were overlooked. As a result, the project failed to account for the realities of the communities' daily lives and their water consumption patterns. Instead of being a joyful and playful solution, the PlayPump became a burden, with some communities reporting that it took hours of continuous use to pump even a fraction of the water they needed.

Maintenance issues compounded the problem. The PlayPumps required regular upkeep and technical support, but these aspects were not sufficiently planned for or funded. Local communities, already struggling with limited resources, were expected to maintain the complex systems without adequate training or support. When the pumps inevitably broke down, they were left unused and abandoned, symbols of a broken promise and failed development.

The Case Foundation withdrew support from the project due to ongoing maintenance issues and user dissatisfaction, leading to the dissolution of PlayPumps International in early 2010.[2]

The PlayPump project illustrates the critical importance of involving the community at every stage of the design process. It highlights that even the most innovative technology can fail if it does not resonate with the lived experiences and needs of the people it is intended to help. Authentic engagement means more than just providing a solution; it involves listening to, learning from, and collaborating with the community to co-create sustainable and meaningful outcomes. Without this, well-meaning projects can not only fail to achieve their goals but also erode trust and leave communities worse off than before.

Engaging with Communities: A Powerful Starting Point

In many instances, I have been amazed by the transformative power of simply engaging with communities.

One project that stands out was for an insurance company looking to improve its services for individuals who experience a loss of mobility following an accident. These individuals often require extensive assistance in their daily lives, provided by caregivers and professionals like nurses. This support, although essential, is costly for the insurance company and can be frustrating for the patients, who typically receive help for only a few hours each day. The company wanted to explore whether a scalable solution, such as a robot, could complement the existing services. At that time, humanoid robots were becoming more prevalent, and the company saw potential in this technology.

Our task was to understand what role a robot could play and to develop a prototype to test its value. We began by conducting customer discovery, interviewing doctors, caregivers, and users. One surprising insight emerged quickly: there were significant similarities between the needs of young adults experiencing temporary loss of autonomy due to an accident and older adults gradually losing mobility over time. This realization shifted the scope and scale of the project, challenging our client's initial assumptions.

To gain a deeper understanding, we immersed ourselves in the environment of those we were designing for by spending a week in a nursing home. This firsthand experience allowed us to truly grasp the daily lives of people with mobility challenges. We also spent time in hospitals and with home-care providers, shadowing them in their daily tasks to better understand the complexities of their work.

The insights we gathered from this on-the-ground research were invaluable. We discovered that the primary needs of people with mobility loss were not functional, such as writing emails or being productive, but emotional. They craved being heard, having meaningful interactions, and witnessing lively scenes around them. Another crucial insight was the importance of supporting not just the patients but also their families and caregivers, who bear the emotional and physical burden of providing care and often experience significant exhaustion and stress.

Moreover, our presence and engagement in the nursing home profoundly impacted the residents. It became a significant event for them, not only because it was a novel and entertaining experience but also because they felt acknowledged and listened to.

One resident, an elderly gentleman who rarely spoke and was known to be quite reserved, surprised everyone when he began engaging in conversations with us about his life and experiences. He shared stories from his youth about his love for music and even gave us advice on navigating life's challenges. The staff told us later that they hadn't seen him this animated in years. Our time there seemed to unlock a part of him that had been dormant, reminding us of the power of simply being present and listening.

Another memorable moment was when a group of residents decided to hold an impromptu talent show to "entertain the researchers," as they put it. One resident, a former dance instructor, took it upon herself to teach us a few dance steps, much to the delight of her peers. The nursing home was filled with laughter and music that day, creating a sense of joy and community that lingered long after we left. It was clear that our presence had sparked a sense of agency and purpose among the residents, allowing them to step out of their daily routines and share a part of themselves they rarely got to express.

There was also an elderly woman who was initially very skeptical of our presence and the purpose of our project. She had seen many initiatives come and go, often with little lasting impact. However, as the week went on and she observed our genuine interest in understanding her and the other residents' lives, her attitude softened. She began to join our conversations and even suggested ways the robot could be more helpful, such as reminding her to take her medication and providing companionship during long afternoons. Her change in perspective highlighted the trust that can be built through sincere engagement and respect.

These experiences underscored the impact that genuine human connection can have, even in the context of a technological project. For the residents, our visits were not just about the potential benefits of the robot; they were a reminder that their voices mattered and that they had valuable insights and stories to share.

For us, it was a powerful lesson in the importance of designing with, rather than for, the community—an approach that can lead to richer, more impactful outcomes for everyone involved.

Inclusive and Equitable Design extends beyond the final product; it encompasses the process of giving a voice to those who are often unheard.

Although community engagement alone is not enough, it offers immense benefits to the design team and all stakeholders involved. It creates a deeper connection to the real needs and desires of the community, leading to more meaningful and effective solutions.

An interesting anecdote from this project occurred when we later programmed the robot to interact with users, have conversations, and respond to their needs. The robot was equipped with a microphone, allowing us to record user interactions and review them afterward. Initially, users seemed underwhelmed by the robot, pointing out its limitations. However, when we analyzed the recordings, we discovered that they had interacted with the robot far more than they had acknowledged and even invited friends over several times to showcase it. These interactions turned out to be a source of joy and entertainment for the users, and the words captured in the recordings highlighted the value and positive impact of these interactions.

This experience reinforced the importance of truly engaging with the community throughout the design process. It is not just about creating a functional product but about understanding and addressing the emotional and social needs of the people we are designing for.

Three Levels of Community Engagement in Design

Engaging with communities during the design process can be approached at three distinct levels: designing for, designing with, and designing by. Each level offers unique benefits and challenges and is relevant in different contexts depending on the goals of the project and the needs of the community. Understanding these nuances is essential for creating impactful, equitable, and inclusive design solutions.

Designing for Communities

Designing for communities involves creating solutions on behalf of a group without their direct involvement in the design process. This approach is often based on research, observations, and assumptions about the needs and preferences of the community. The design team takes on the role of an external expert, interpreting the needs and problems of the community to develop appropriate solutions.

- Pros:

  - **Speed and efficiency:** Because the design team operates independently, the process can be faster and more efficient, without the complexities of coordinating with a broader group.

  - **Expertise-driven solutions:** Designers can leverage their expertise, knowledge, and resources to create technically advanced and innovative solutions that the community might not have the capacity to develop on its own.

  - **Ideal for initial interventions:** This approach is useful when there is an urgent need for a solution and the community lacks the resources or expertise to address the issue themselves.

- Cons:

  - **Lack of community input:** Without direct involvement from the community, the solutions may miss critical insights and fail to fully address the unique needs and context of the users.

  - **Risk of misalignment:** The resulting design may not resonate with the community's values, culture, or lived experiences, leading to lower adoption or even rejection.

  - **Potential dependency:** This approach can create a sense of dependency, where the community relies on external parties for solutions rather than being empowered to develop their own.

- When relevant:

  - **Emergency situations:** When immediate solutions are required, such as in disaster relief or public health crises.

  - **Initial prototyping:** In the early stages of a project, to quickly develop and test concepts before involving the community in the iterative process.

  - **Example:** During a public health crisis, a tech company urgently developed a mobile health app for rural communities in developing countries. With the disease spreading faster than resources could be mobilized, the communities needed a digital tool to quickly provide critical information. Relying on secondary research and expert input, the company created the first iteration of the app, knowing it might miss key cultural nuances but prioritizing immediate functionality to save lives.

This rapid deployment acted as a vital stopgap, addressing the most urgent needs until the company could gather direct user feedback to refine the app. Although not perfect, this initial version was essential in the face of the urgent crisis.

Designing with Communities

Designing with communities involves actively involving members of the community in the design process as collaborators. This approach values the input and feedback of the community, fostering a co-creation environment where designers and community members work together to develop solutions.

- ▪ Pros:
  - ▪ **Rich insights and relevance:** By engaging directly with the community, designers gain valuable, real-world insights that lead to more relevant and contextually appropriate solutions.
  - ▪ **Increased buy-in and ownership:** When community members participate in the design process, they are more likely to feel a sense of ownership and commitment to the solution, increasing its chances of success.
  - ▪ **Capacity building:** This approach helps build skills and knowledge within the community, empowering them to address their own challenges in the future.
- ▪ Cons:
  - ▪ **Time-consuming:** Co-creation requires significant time and effort to build trust, facilitate workshops, and incorporate diverse perspectives, which can slow down the design process.
  - ▪ **Complex coordination:** Managing collaboration between designers and community members with varying levels of expertise and differing viewpoints can be challenging.
  - ▪ **Scope of influence:** The community's involvement is usually limited to the scope of the project defined by the designers, which may not fully address all the community's needs.
- ▪ When relevant:
  - ▪ **Community-centric projects:** When the goal is to create solutions that are deeply embedded in the community's cultural and social context, such as urban planning or educational programs.
  - ▪ **Iterative development:** When a project benefits from continuous feedback and iterative improvements based on real user experiences and suggestions.
  - ▪ **Example:** A city government works with local residents to redesign a public park, conducting workshops and feedback sessions to ensure that the space meets the needs of the community.

Designing by Communities

Designing by communities involves empowering community members to take the lead in the design process. Here, designers act as facilitators or supporters, providing resources, tools, and guidance to help the community create their own solutions. The community holds the primary decision-making power and is actively engaged in all stages of the design process.

- ▪ Pros:
  - ▪ **Empowerment and self-reliance:** This approach fosters a strong sense of empowerment and ownership, as the community drives the process and outcome, building confidence and capacity.
  - ▪ **Deep cultural relevance:** Solutions are likely to be more culturally appropriate and sustainable because they are created by those who fully understand the community's context and challenges.
  - ▪ **Long-term impact:** By equipping communities with the skills and resources to design their own solutions, this approach can lead to more sustainable and long-lasting outcomes.

- ▪ Cons:
  - ▪ **Resource intensive:** Providing the necessary training, resources, and support can be costly and time-consuming.
  - ▪ **Risk of limited expertise:** If the community lacks certain technical skills or knowledge, the solutions may not be as sophisticated or effective as those designed with external expertise.
  - ▪ **Potential for internal conflict:** Engaging the entire community in the decision-making process can sometimes lead to disagreements or conflicts, which can stall or derail the project.

- ▪ When relevant:
  - ▪ **Grassroots initiatives:** When the goal is to address issues that are deeply rooted in the community and require local knowledge and leadership to solve, such as community health programs or local economic development.
  - ▪ **Long-term sustainability:** When it's crucial to build long-term capacity and self-sufficiency within the community.
  - ▪ **Example:** A rural community takes the lead in designing and implementing a clean water system, with support from an NGO that provides technical training and resources, ensuring that the community has the skills to maintain the system long after the NGO has left.

In summary, choosing between designing for, designing with, and designing depends on the project's goals, timeline, and the community's capacity and needs. Understanding when to apply each approach is key to achieving meaningful and sustainable outcomes that truly benefit the communities involved.

Different Paths to Inclusion: Stories of Designing for, with, and by Communities

Let's explore different stories—from my own experiences and from those of renowned designers and industry leaders—to illustrate how these various approaches take shape and what the journey entails.

Designing for Communities: The World-Renowned OXO Peeler

In the early 1990s, the kitchenware industry saw a quiet but groundbreaking revolution with the launch of OXO's Good Grips line. It all began with a vegetable peeler (see Figure 6.2), a common tool that millions of people struggled to use daily, especially those with arthritis or limited dexterity. Sam Farber, OXO's founder, was one of the few who recognized this problem. His inspiration came from a deeply personal place: his wife, Betsey, who had arthritis and found traditional kitchen tools uncomfortable and painful to use.

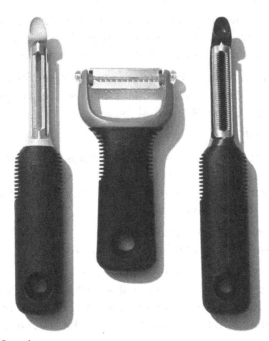

Figure 6.2: The OXO peeler

Farber's response to this challenge was rooted in empathy but also in a pragmatic "designing for" approach. In collaboration with the renowned design firm Smart Design, OXO set out to create products that addressed well-documented and widely understood challenges faced by individuals with limited hand strength. The focus was not on co-creation or direct user involvement but on leveraging existing research, expert insights, and ergonomic principles to design better tools.

The first product, the OXO Good Grips vegetable peeler, debuted in 1990. It was a striking departure from the sharp, thin, and uncomfortable designs that dominated the market. Featuring a thick, rubberized handle with ridges for better grip and control, the peeler immediately stood out. The design prioritized comfort and usability, reducing the strain on users' hands. Notably, it was tested rigorously by Farber and the design team but not extensively co-designed with end users—a hallmark of the "designing for" methodology.

The gamble paid off. Within its first year, the peeler became a bestseller, earning praise not only from the arthritis community but also from the general public. The design, although initially intended for a specific user group, resonated with a broad audience. OXO sold over 500,000 peelers in the first 12 months and quickly expanded the Good Grips line to include other kitchen tools, such as can openers, spatulas, and whisks, all following the same ergonomic principles.

The success of Good Grips highlighted the strengths of the "designing for" approach. By focusing on well-documented needs and applying thoughtful design principles, OXO was able to create a product that solved a specific problem while appealing to a wider market. This approach allowed for rapid development and scalability, which is critical for gaining an early foothold in a competitive market.

However, the approach was not without its limitations. Without direct input from end users, there was always a risk of overlooking subtle preferences or unique needs that could have further enhanced the design. For example, although the rubberized handle was widely praised, some users later expressed that additional grip textures could have improved usability under wet conditions. These insights might have been captured earlier with a "designing with" or "designing by" approach.

The Good Grips line has since grown to include hundreds of products, and OXO has become a household name synonymous with thoughtful, inclusive design. The brand has earned numerous accolades, including a place in the Museum of Modern Art's collection, and continues to generate millions in revenue annually. The story of OXO's Good Grips exemplifies the power of "designing for" as a strategy—one that uses research, empathy, and expert insights to address specific challenges while creating products that resonate universally.[3, 4]

Designing with Communities: A Neurodiverse-Friendly Productivity Suite

Contrast that with our work around the flagship productivity software of a leading technology company to make the suite more accessible for neurodiverse individuals. This project, led by a group of four of my students at Berkeley, couldn't have been more different. From the beginning, we knew that designing with the community was essential. We weren't just trying to make minor adjustments; we were rethinking the user experience for a group of people who often struggle with traditional software interfaces.

We organized workshops and focus groups with neurodiverse users, where we didn't just listen—we watched. We observed how they interacted with the software, where they got stuck, and what frustrated them. One participant, who had ADHD, described the traditional ribbon interface as a "mental minefield." Too many options, too many distractions. Another user on the autism spectrum shared how the barrage of notifications felt overwhelming, almost like being shouted at by the software.

These insights were invaluable and shaped every step of our design process. We introduced a simplified ribbon view, reduced visual clutter, and created customizable settings that allowed users to tailor the software to their unique needs. By designing with the community, we didn't just build a more inclusive product; we built one that felt personally tailored to the individuals who needed it most. This approach was right for the project because it allowed us to address very specific personal challenges that couldn't have been identified without direct user involvement.

Although "designing with" communities fosters collaboration and creates solutions that deeply resonate with community needs, it can be a time-intensive process. Building trust, conducting workshops, and incorporating diverse perspectives require significant resources and often extend project timelines. This can pose challenges in scenarios requiring quick action, such as disaster relief or urgent public health responses, where a slower, inclusive approach may delay critical outcomes.

Additionally, coordinating input from diverse community members can be complex, especially when priorities conflict. Balancing these differing viewpoints while maintaining a cohesive design vision is no small feat and can lead to decision-making bottlenecks. Moreover, community involvement is often limited to the scope of the project as defined by designers, leaving broader concerns unaddressed. These challenges highlight the need for careful planning and transparency to ensure that the benefits of co-creation outweigh its limitations.

In this specific project, "designing with" was the most appropriate approach due to the unique and complex context of the challenge. The relative novelty of the problem, coupled with limited existing literature or established frameworks, made it difficult to rely solely on prior research or external expertise. Additionally, the intricacies of the community's lived experiences presented challenges that were nearly impossible to fully grasp without direct involvement. The hardships and specific needs faced by the users required a level of empathy and insight that could only be achieved through active collaboration. By co-creating with the community, the project not only addressed these gaps in understanding but also ensured that the final solution was both relevant and effective, reflecting the real-world realities and priorities of those it sought to support.

Designing by Communities: Empowering Small Business Owners

In one studio project, we were tasked with developing a digital tool to support small business owners. Our small team of designers and consultants aimed to adapt existing enterprise software into a customizable solution for independent business managers. We needed to create some mockups and offer strategic directions to the editor.

The challenge was immediately apparent: small businesses are remarkably diverse. The operational needs of a family-owned restaurant are fundamentally different from those of a boutique retail store or a freelance graphic designer. No single, one-size-fits-all solution could adequately address this variety.

Recognizing this, we embraced a "designing by" approach, empowering users to take the lead in shaping the tool to fit their unique requirements. Instead of dictating specific features, we built an open platform with customizable workflows and automation capabilities. Users could create their own setups, share insights in community forums, and suggest new features. This user-driven model fostered an organic evolution of the tool. A small bakery owner, for instance, designed an inventory management workflow that became a template followed by dozens of other users. A freelance photographer created a project-tracking template that quickly gained popularity among creative professionals. The platform became a living, evolving ecosystem shaped by the ingenuity and needs of its users.

However, this "designing by" approach also came with challenges. It required significant resources to support the community with robust training materials, technical assistance, and forum moderation. Some users struggled with the steep learning curve, and the open-ended nature of the platform meant that not all solutions were equally effective or efficient. Additionally, relying on the community's input could lead to uneven adoption of best practices, as not every user's contribution was well-suited for broader use.

Despite these challenges, the "designing by" approach was the right methodology for this project. The relative novelty of the problem—a digital tool tailored to the vast diversity of small businesses—meant there was little precedent or existing literature to guide us. The sheer variety of needs made it impossible to predefine all potential use cases, and stepping fully into the shoes of the end users would have been a monumental task. By empowering the users themselves, we harnessed their firsthand knowledge and ingenuity, resulting in a tool that was versatile, adaptive, and continuously refined by its community.

Each of these projects shows how choosing the right level of engagement can transform the impact of a design. Whether we're designing for, with, or by a community, it's about understanding what's needed and being flexible enough to meet those needs in the most effective way possible.

Bringing Voices to the Table: The Power and Challenges of True Representation

The most straightforward and impactful way to truly understand and amplify the voices of your target audience is to bring them into the conversation from the very start. Having a diverse team is not just beneficial; it's essential. It has been shown to lead to better outcomes, greater resilience, and even improved financial performance. But for those of us working on Equitable Design projects, it goes beyond just ticking the diversity box. Having team members whose backgrounds and experiences closely align with those of our target audience is invaluable. It allows us to have more authentic conversations, bridges gaps in understanding and helps us take the right approach to solving complex issues. It also enables us to connect more deeply with the wider community we are trying to serve.

Although it may seem like an obvious strategy, it's worth reiterating because it is so crucial and often overlooked.

I recall a student project we undertook with the corporate venture arm of a leading consulting firm focused on creating more opportunities for Black entrepreneurs. The students chose their projects, and I created teams of four or five people, aiming for the best diversity in the team but respecting their team preferences; they then worked together for the semester, acting as consultants for the partner. We were fortunate to have two Black entrepreneurs on our team. Their contributions were more than just insightful—they were transformative. They provided firsthand experiences and perspectives that made our analysis more accurate and our solutions more relevant. Their presence also gave us immediate access to a larger network, allowing us to test our ideas more effectively and refine them based on direct feedback.

In another instance, when we were working on a consulting project related to disability, we had the guidance of a close friend of mine who had been living in a wheelchair for 40 years. Although he was an outside contributor supporting our team essentially because we were friends, his mentorship was a game-changer. He helped us navigate the nuances of discussing disabilities, understand critical emotions and experiences, and approach the subject with the sensitivity and depth it deserved. His input wasn't just valuable—it fundamentally shaped the direction and impact of the project.

Whether these individuals are full-time team members, advisors, or part-time contributors or serve in other roles, their involvement can be incredibly beneficial, provided they are integrated under the right conditions. Unfortunately, this isn't always the case.

We've also faced situations where the setup was less than ideal. There were times when some team members were not as open-minded or constructive as they needed to be, which led to tensions. This is especially challenging in contexts where team members are not just contributors but also representatives of the communities we are designing for. This dual role can create an imbalance, making it difficult for them to fully participate in the team dynamic without feeling the weight of representing an entire community.

One particularly challenging experience occurred during a university project focused on housing and urban safety. The team comprised a group of enthusiastic freshmen brimming with ideas and a graduate student who had personally experienced homelessness and struggled with mental health challenges. Their task was to develop solutions for individuals living in cars or vans. The diverse perspectives on the team initially seemed like a strength but soon proved to be a double-edged sword.

The freshmen were eager to explore innovative solutions and excited about the potential for creative design and technology. In contrast, the graduate student, drawing from lived experience, found many of these ideas disconnected from the harsh realities faced by people living in such precarious conditions. The team quickly hit a wall as they struggled to bridge the gap between theory and reality. Each side felt passionately about their viewpoint, and neither was willing to fully listen or compromise.

As the project progressed, the tension escalated. The freshmen felt frustrated that their enthusiasm and ideas were being dismissed, and the graduate student felt misunderstood and sidelined, believing the team's approach lacked empathy and practicality. It became increasingly clear that they weren't just working from different perspectives—they were speaking entirely different languages.

Despite our attempts to facilitate discussions and find common ground, the divide grew wider. The inability to reconcile these differences hindered progress, and the project stalled. Eventually, we had to make the difficult decision to split them into subteams. It was a decision that felt more like a stopgap than a solution—a necessary compromise in the face of irreconcilable differences. It served as a stark reminder that although diversity of thought is invaluable, without mutual understanding and the willingness to truly engage with each other's experiences, it can also become a significant barrier to collaboration and progress.

Looking back, we should have provided better tools and techniques for team bonding and decision-making and perhaps included more team members who could mediate between these two extremes. This experience taught us that diversity is not just about bringing one person from the target audience or a minority group onto the team. It's about creating a truly balanced and diverse group where multiple perspectives can coexist and contribute meaningfully to the project.

In the end, it's not enough to simply invite diverse voices to the table; we must also ensure that the environment allows for genuine collaboration and that all perspectives are heard and valued. Only then can we achieve the harmony and effectiveness needed for impactful, Equitable Design.

Demystifying Community Outreach: Embracing Vulnerability to Build Trust

I've often noticed that both students and corporate leaders hesitate to engage directly with their target communities. They worry about not having access, being seen as intrusive, or feeling awkward discussing sensitive topics. Some people just don't feel comfortable talking to strangers. It's a natural reaction.

What I usually tell them is that if you go in with good intentions and a genuine purpose—to help solve a real problem—you'll likely be welcomed. It's all about being upfront about why you're there and what you're trying to achieve. You've got to show people that you respect their time and perspectives and that you're not just there to take what you need and leave.

But there's more to it than just being honest and clear. A key part of building trust is showing vulnerability. Admitting that you don't have all the answers, that you're not the expert on their lives, can go a long way.

I mentioned vulnerability in an earlier chapter and will return to it later, emphasizing how integral it is to the backbone of the Inclusive Design mindset and approach.

When I approach a community, I often start by saying that I'm there to learn, that I have my own struggles and failures, and that I need their help to understand what's really going on. This kind of openness creates a space where people feel more comfortable sharing their own experiences because they see that I'm not there to judge or impose but to listen and collaborate.

Of course, it's not always that simple. In some places, especially those with high crime rates or where people feel constantly scrutinized, there's a natural wariness of outsiders. I've worked in areas like the Middle East, where people live under regimes that use spies and other forms of surveillance. In these environments, people are understandably guarded and worried about who they're talking to and why.

Then there are situations closer to home where people have had to bend the rules to get by—maybe they've found ways to make money that aren't exactly above board, or they're trying to save on taxes. They won't be eager to talk to someone who might expose them, even unintentionally.

And it's not just about laws or politics. Cultural and family dynamics play a big role, too. In some communities, speaking with women can be tricky because men are always watching. I've been in situations where families were afraid to talk about a disabled member because they were worried it could jeopardize the welfare benefits they relied on. In all these cases, trust is the key.

Building that trust takes more than just explaining your project. It's about how you carry yourself and how you interact with people. Being transparent about why you're there helps, but so does showing vulnerability. I often start by admitting that I don't know much about their lives and that I'm here to learn.

I share my own struggles and failures, which puts us on a more even footing and helps them see me as a person, not just someone from the outside trying to get information.

These small gestures can make a huge difference. They can help you connect with people who might otherwise be reluctant to open up. And when that connection happens, you'll often find that people are far more willing to share than you'd expect. That experience can be incredibly rewarding and will encourage you to do more of this kind of outreach.

Not every attempt will be successful. There will be tough conversations and times when people don't want to talk. That's okay. It's rarely about you or your project specifically; it's often about deeper frustrations or distrust toward the system as a whole. Just keep at it, stay honest and open, and remember why you're doing this in the first place. Vulnerability isn't a weakness; it's a way to build real, meaningful connections with the communities you're trying to serve.

The Prototyping Strategy: Making Every Moment Count

Engaging with communities isn't only about showing up and talking to potential users; it's also about using that time effectively to make real progress on the project. This is where a prototyping strategy, especially rapid prototyping, becomes crucial.

Rapid prototyping is all about turning an idea into a tangible form as quickly as possible, whether that's a simple sketch, a rough model, or a basic digital mockup. The goal is to create something real that people can interact with, critique, and help improve. The concept might seem counterintuitive at first—why rush to build something half-baked? But the key is to maximize learning with minimal effort.

By spending less time perfecting your initial prototype, you can create multiple iterations quickly and gather feedback rapidly. This approach also reduces your emotional attachment to the prototype. When you've poured weeks or months into a single design, it's hard to let go and make changes based on feedback. But if you've only spent a few hours, you're more open to altering and refining the concept based on what you learn from the community.

Another benefit of quick, rough prototypes is that they invite criticism. A polished product can be intimidating for people to critique—they don't want to hurt your feelings or feel like they're destroying your hard work. But a rough mockup or simple sketches encourages honest feedback because it's clear that this is just a starting point, not a finished product.

One of the most memorable examples of this took place in 2017. The CEO of a major national bank in Europe faced criticism from journalists questioning the bank's claim of being "the bank for everyone" while failing to provide evidence to back it up. After a challenging on-air confrontation, the CEO turned to his close collaborators, urging them to develop product innovations that could demonstrate the bank's commitment to welcoming unbanked populations. The responsibility eventually fell to me, as the bank outsourced the creative process to my team.

Working with two designer colleagues, we felt the weight of the pressure. Building an app was a non-negotiable requirement from our client, but instead of diving straight into digital design, we began with simple paper mockups of the app's interface. We sketched the basic screens and functionalities by hand, intentionally keeping everything minimal and focused. Our goal was to prioritize the service we were offering rather than getting distracted by aesthetics or technical details at such an early stage.

We then took these paper prototypes to community centers and handed them out to our target audience, along with some pens and markers. Almost immediately, people began scribbling their ideas onto the paper—adding buttons, suggesting new screens, and even reimagining the layout. They felt free to tear apart our initial concept because it was just a bunch of drawings, not a polished product. They knew we had plenty of copies, so they didn't worry about "ruining" anything.

This process was incredibly valuable. Not only did we get a wealth of insights and suggestions that we wouldn't have thought of on our own, but the people felt empowered—they had a hand in shaping the product from the ground up. We moved from merely designing for the community to designing with them and, ultimately, letting them take the lead, designing by the community.

By the end of the process, we had something much more aligned with the needs and desires of the people who would actually be using the app. This approach, which started with a few sheets of paper, turned into a meaningful co-design experience that brought everyone's voice to the table.

It's important to note that rapid prototyping is not exclusive to Inclusive or Equitable Design; it's a well-established technique in design thinking and product development across industries. However, what makes it particularly impactful in this context is how it has been adapted to prioritize inclusivity and empowerment. For example, although rapid prototyping often focuses on speed and efficiency, in Inclusive Design, it also becomes a means to actively involve underrepresented voices in the creative process. By using tools like paper mockups or other accessible media, the approach ensures that people from diverse backgrounds—who may lack technical expertise or familiarity with traditional design processes—can still engage fully and meaningfully. This adaptation transforms prototyping into a collaborative and democratized process, enabling designers to learn not just *about* the community but *with* the community. In doing so, it creates solutions that are innovative and also deeply rooted in the lived experiences of the people they aim to serve.

Seeing Through Their Eyes: Using Photo-Elicitation for Deeper Community Insight

There are times when no matter how much preparation we do, we are too removed from the target audience. The cultural or language barriers may be too significant, making it challenging to connect and gather meaningful insights through traditional methods. In these situations, I've found that certain techniques can bridge the gap and provide a deeper understanding of the community's needs and perspectives.

One such technique is called *photo-elicitation*. This method involves using photographs to evoke responses and facilitate conversation. Instead of relying solely on words, which can sometimes be limiting or misinterpreted, photo-elicitation leverages the power of imagery to open up dialogue. It's a way of tapping into the visual language that often transcends verbal communication barriers.

For example, when working on a research project with students and health policy experts to improve access to healthcare in underserved communities, especially immigrants, we found it difficult to get beyond surface-level responses. People were hesitant to share their thoughts openly, either because they didn't feel comfortable discussing their struggles with outsiders or because they lacked the vocabulary to express their experiences. So, we decided to try something different: we provided disposable cameras to several community members and asked them to take photos of their daily routines, focusing on anything related to health and well-being.

When we later sat down with these individuals to discuss their photos, the images sparked stories and insights that wouldn't have emerged otherwise. For example, a photo of a long line outside a community clinic led to a conversation about the frustration of waiting hours for basic care. A picture of a neighbor helping someone with a disability navigate a set of broken stairs highlighted the community's resilience and mutual support but also pointed to the need for better infrastructure.

Photo-elicitation allows people to frame their narratives: to show what matters most to them through their own lens. It's a technique that honors their perspective, giving them control over how their stories are told. This is particularly relevant in the context of Equitable Design, where the goal is to create solutions that resonate with and truly benefit the community. By involving people in the storytelling process, we're not just collecting data; we're empowering them to share their lived experiences in a way that feels natural and authentic.

The value of this approach goes beyond overcoming language or cultural barriers. It's about recognizing that people are the experts in their own lives. They may not always have the technical language to describe their needs, but through images, they can convey complex emotions, challenges, and desires. It shifts the dynamic from one where we, as designers, are the "question-askers" to one where the community members are co-creators of the narrative.

Another example of using photo-elicitation was during a research project I overviewed with a department of the French government that was looking at the future of disability, with a focus on improving urban spaces for people with disabilities. Rather than conducting standard interviews, we asked participants to photograph places in their neighborhoods that they found challenging to navigate. The resulting images were powerful: narrow sidewalks, lack of ramps, and inaccessible public transportation options. These visual stories provided a more visceral understanding of the everyday obstacles faced by people with disabilities and helped us identify priority areas for design intervention.

Photo-elicitation is a well-established method widely used in anthropology, sociology, and design to evoke deeper, more authentic responses from participants. However, in the context of Equitable and Inclusive Design, its relevance becomes even more pronounced. This methodology shifts the power dynamic, allowing participants—especially those from underserved or marginalized communities—to control the narrative and highlight what truly matters to them. By centering their perspectives, photo-elicitation not only gathers data but also amplifies voices that are often excluded from traditional design processes.

In Equitable Design projects, the stakes are higher, and the contexts are often more complex than in standard design initiatives. The lived experiences of participants in these contexts are deeply personal and shaped by systemic inequities, cultural nuances, and emotional challenges. Traditional research methods might inadvertently flatten these complexities or miss them altogether. Photo-elicitation, in contrast, captures the richness and depth of these experiences, providing a holistic understanding that is crucial for designing solutions that genuinely resonate with the target community.

By adapting photo-elicitation to these contexts, we're not just employing a tool but redefining its purpose. It becomes a bridge to empathy, a means to uncover hidden truths, and a way to foster a collaborative design process that prioritizes dignity and inclusivity. This technique ensures that the voices of those who live at the intersection of design challenges are heard, respected, and central to shaping the outcomes. In doing so, photo-elicitation transforms from a research tool into a cornerstone of Equitable Design practice.

From Overlooked Details to Design Inspiration: Immersing in User Realities

Often, the people we design for have developed their own ways of navigating obstacles, so they may not immediately recognize certain pain points as exclusionary. They adapt, making the best of what's available, and as a result, they may not always feel left out of an experience. But in my experience, there are always moments and situations where the exclusion is both real and problematic—sometimes it just takes deeper engagement to uncover them.

Take the time we were working on an authentication service for people with visual impairments. In initial interviews, participants told us they had found ways to manage and didn't think this app was harder to use than others, especially because it had some accessibility features. On the surface, everything seemed fine. But as we spent more time with the contributors, we had a moment that completely shifted our understanding. One day, while hanging out at the library, we saw a person we had interviewed earlier sitting nearby. In the quiet of the library, when they had to log in to their account, we all heard their accessibility tool loudly announce their password. This was a striking moment. It wasn't something the beneficiary had brought up directly, but we could see—and hear—how this created a real privacy issue in public spaces. This insight reshaped our entire approach. Moving forward, our focus became developing a solution that didn't rely on vocal interactions for authentication.

Another example involved working with visually impaired people on their experience with video streaming services. Initially, we were focused on making the actual movie-watching experience more seamless, but during our interviews, we couldn't identify any major pain points beyond what had already been addressed by existing features. After wrapping up one session, we went back to their place to thank them, where we met their daughter. During a casual conversation, the father and daughter began talking about the process of selecting a movie—a task we hadn't considered. They described how frustrating and tedious it was for them to pick something to watch together, given the lack of accessible options for browsing titles. This shared story brought an emotional depth to our understanding of the problem. We realized the most significant pain point wasn't the movie-watching itself but navigating the selection process—a challenge we had completely overlooked. This led us to refocus on redesigning the browsing experience, with that father–daughter story always in the back of our minds as we worked on the solution.

A third example that stands out came from research on notebooks for children with neurodivergence. We had already gathered solid insights from interviews with parents, but the real breakthrough happened during a home visit. As we spoke with one mother, she casually mentioned her son's relationship with books. He loved them, she said, but reading and writing were tough for him. What really caught our attention was when she said "He's trained himself to listen at *2x speed*—we can't even understand it, but he comprehends super-fast speech." This was a game-changing detail that shifted how we looked at the problem. From that point, we began exploring text-to-speech solutions that catered to his specific way of processing information, and this opened up entirely new possibilities for the project.

These examples show the importance of immersing yourself in the lives of the people you're designing for. It's not just about conducting interviews or gathering feedback—it's about being present, paying attention to the fringe cases, and listening for the unexpected stories that reveal the most critical insights. Often, those hidden or overlooked moments hold the key to truly inclusive design.

Accessing Inaccessible Communities

Sometimes we have to acknowledge that reaching the audience we aim to serve is incredibly challenging. It could be due to geographical distance, project limitations, or even safety concerns. But even in these situations, there are still ways to build bridges, and making the effort is always worth it because of the value these connections bring.

Let me take you back to a project I worked on years ago. It was during the height of the civil war in Syria, and thousands of people were fleeing the country, with many crossing into neighboring Lebanon. The situation was desperate. Families left everything behind, traveling with only what they could carry to save their lives. But their suffering didn't end once they crossed the border.

Lebanon, a country I deeply admire for its rich culture, incredible history, and unforgettable hospitality, faced its own challenges. It's a country of remarkable complexity: one-third Sunni Muslim, one-third Shia Muslim, and one-third Christian, each with deep historical roots. This delicate balance, combined with its geography—wedged between Israel, Syria, and the Mediterranean Sea—meant that Lebanese authorities were understandably cautious about any disruptions to the fragile peace.

With this backdrop, the influx of Syrian refugees was met with a mix of welcome and hesitation. Although Lebanon opened its doors to millions of refugees, there were strict policies in place to ensure that these settlements wouldn't turn into permanent cities, which could create instability in areas lacking infrastructure. One of the rules was that shelters for refugees had to remain temporary, meaning they were made from basic scrap materials—wood beams and tarps.

Here's where our team came in. We were part of a unique consortium that brought together three distinct partners. On one side was a large global NGO, a world leader in providing basic sanitary infrastructure for refugees and endangered populations in zones of distress. The second partner was a top-five global construction company eager to share its resources and expertise to contribute to a major humanitarian cause. The final partner was my design studio, which specialized in assembling workforces of designers, students, and experts to tackle complex problems in a short amount of time. Together, we had just a few months to collaboratively develop scalable solutions to support Syrian refugees in the camps.

The conditions in these makeshift shelters were dire, particularly during the winter months. A lot of people don't realize how cold it gets in Lebanon. The country's mountains see regular snowfall, and the shelters were ill-equipped to handle these freezing temperatures. Our mission was to design shelters that could improve living conditions, particularly addressing the biting cold that people were enduring.

We had to get creative in finding ways to gather information and connect with the community remotely. Here are four strategies we used to engage with our end beneficiaries and collect the data we needed.

Ask Local Experts

First, we focused on reaching out to experts and witnesses—people who had experienced the conditions in Lebanon firsthand but were more accessible to us. We connected with NGOs, humanitarian workers, and volunteers who lived in our region but traveled frequently to Lebanon. These individuals had a deep understanding of the situation on the ground and could provide valuable, first-hand accounts of what was happening. They shared photos, videos, and stories, giving us a clear picture of the realities facing the refugees.

What made these intermediaries particularly valuable was their familiarity with both cultures. They understood the context in Lebanon, but they also knew how to communicate those insights in a way that made sense to us, bridging the cultural and knowledge gaps that could have otherwise held us back. Their ability to translate their experiences into actionable insights was crucial to our project.

Beyond the data they provided, these experts became sources of inspiration for our team. Their dedication and commitment to helping communities in need fueled our determination. Hearing their stories and seeing their work on the ground gave us a sense of purpose and drive that made our project feel even more meaningful. The connection was mutual—their appreciation for our efforts to contribute was equally fulfilling, creating a shared sense of purpose that energized both sides.

Engaging with experts and people who work directly with the communities you're designing for is always a productive strategy. Not only do they provide essential insights, but they also offer motivation and guidance, reminding you of the impact your work can have.

Leverage Netnography to Uncover Unique Insights

One highly effective technique we used was *netnography*—a method of conducting ethnographic research online. Unlike simply searching for articles or media coverage about the topic at hand, netnography digs deeper, exploring how real people experience and share their lives through digital spaces like blogs, forums, and social media.

Netnography allows you to see the world through the eyes of real users by following their online activity, studying their posts, and paying close attention to the details they choose to share. It's a way of capturing everyday experiences in a much more authentic and granular way. You can even engage directly with people in these spaces if it feels appropriate, opening up the possibility for valuable conversations and feedback.

These platforms are rich sources of information because they offer unfiltered, real-time insights into people's thoughts, behaviors, and environments. People often feel freer to speak openly and honestly when they're behind a screen, especially in forums where anonymity encourages raw and candid discussions.

In the context of our work in Lebanon, netnography provided a window into a highly connected population, which helped break down barriers and challenge some of the preconceived ideas we initially had. It was eye-opening to see how these communities were living through their posts and interactions, giving us a more nuanced understanding of the challenges they faced.

Explore Analogous Situations

Another way we gathered valuable data was by exploring analogous situations. Sadly, the plight of refugees isn't new, and we were able to draw from the experiences of others who had lived through similar hardships. We connected with migrants who had previously lived in refugee camps and temporary shelters across Europe. They shared their stories of survival, describing how they managed to cope with harsh winters in makeshift shelters. Their stories were filled with creative solutions and workarounds they'd developed under incredibly tough conditions.

One powerful insight we gained was that the real struggle wasn't just the cold but the humidity that often accompanied it. When it rained or snowed, everything inside the shelters got wet, and as they explained, that dampness made the cold penetrate straight to the bone. They shared practical tips about how they managed to stay dry and what materials worked best to fight off the cold in such conditions.

We also spoke with mountain climbers and extreme hikers, whose situations, although different in motivation, offered us valuable lessons. These experts, driven by a passion for performance, were full of fascinating ideas about how to design lightweight, durable solutions that could withstand extreme weather. They told us about the best materials to use, how to maximize heat production with limited resources, and how to adjust to changing weather conditions. Their insights were a huge source of inspiration for our designs, adding another layer of depth to our understanding and helping us create more effective solutions for the refugees.

Try to Create a Living Lab

One final technique we used, which ties back to some of the earlier strategies, was replicating the setup of our stakeholders. For this project, we ordered wood beams and tarps and, based on descriptions and plans we'd gathered from our various discussions, attempted to re-create the shelters ourselves. It became obvious pretty quickly that we weren't up to the task—our construction skills were lacking! But that in itself was a powerful learning experience. It added another layer of empathy and understanding to the challenges faced by the people living in these conditions.

We spent several days and nights in our makeshift shelter, and the experience revealed new dimensions to the situation. For example, using a woodfire cookstove in a tarp shelter proved to be far from straightforward. We had to balance the trade-off between thermal insulation and proper smoke ventilation—something we'd only really understood in theory before but now grasped viscerally.

In every Equitable Design project where we've managed to replicate real-life conditions and create a kind of living lab, the insights have come faster and more clearly. This hands-on approach helped us iterate more effectively, as we could feel firsthand the compromises and challenges our users face daily. It brought us much closer to designing meaningful, practical solutions.

Ethical Considerations and Power Dynamics in Equitable Design

When working on Equitable Design projects, ethical considerations and power dynamics are always at play. The way we engage with communities—how we ask questions, whose voices we amplify, and how we handle feedback—can either foster trust or exacerbate the very problems we aim to address. Designing for marginalized communities requires us to be acutely aware of the power imbalances that exist between designers and the people we hope to serve.

Equitable Design isn't just about making a product or service accessible; it's about shifting the balance of power so that communities are not only represented but also actively engaged and involved in decision-making. This requires a thoughtful approach to how we handle consent, participation, and compensation and how we ensure that the process itself is fair and empowering.

Transparency and Building Trust

One of the most effective ways to address power dynamics is through transparency. When communities understand the goals of a project and what's being asked of them, they are far more likely to engage meaningfully. In one particularly memorable project focused on equitable real estate practices, transparency was not just a strategy but a necessity.

This consulting project I led with a couple of colleagues aimed to explore housing solutions for low-income renters, a group often burdened by mistrust of institutions due to systemic inequities and exploitative practices. Early on, we noticed significant resistance from potential participants, many of whom were wary of our intentions. Among them was Ana, a participant who initially avoided all attempts to engage with her.

Ana had recently been through the grueling process of defending herself in court over her over-indebtedness, an experience that had left her emotionally exhausted and distrustful of any initiative she feared might expose her financial struggles further. She consistently turned down our invitations to interviews, avoiding any discussion related to real estate or financial matters. For Ana, our project seemed like another intrusion into an already overwhelming and painful part of her life.

Understanding the sensitivity of the situation, we decided to focus on complete transparency to build trust. We created an accessible document and a short, clear video that outlined who we were, the project's goals, and, most importantly, what participation entailed—and what it didn't. We reassured participants that our work was entirely voluntary, confidential, and independent of any financial institutions or processes that might affect their credit, housing status, or legal situation. We even invited a trusted community advocate to review our materials and vouch for our intentions. We used simple wording and a FAQ style to make sure the message was delivered in the best way possible.

It was a turning point. Ana later shared that after reviewing the materials and speaking with the advocate, she felt reassured that we weren't there to exploit her story but to genuinely learn from her experiences. She agreed to an interview. What followed was extraordinary: not only did she share practical insights about the challenges of navigating the real estate market as someone with financial struggles, but she also opened up about the emotional toll of eviction notices and the despair of facing judicial proceedings. She described the profound fear and shame she felt when receiving official letters, the sleepless nights spent worrying about her children, and the small but meaningful moments of solidarity from neighbors who had faced similar struggles.

Ana's willingness to open up transformed our understanding of the problem. She didn't just provide data; she painted a vivid picture of what it feels like to live under the constant shadow of financial instability. Her story allowed us to see beyond the numbers and policies and into the lived reality of those most affected.

By prioritizing transparency and addressing participants' fears head-on, we gained access to critical insights and also created a space where participants like Ana felt empowered to share their truths. This approach underscored a powerful lesson: transparency is not just about providing information—it's about dismantling barriers of mistrust and creating conditions for authentic, human-centered collaboration.

Inclusion Beyond the Loudest Voices

Another crucial aspect of Equitable Design is ensuring that all voices are heard, not just the most outspoken. In one project aimed at improving banking services for underbanked individuals, we were conscious that the most vocal community members aren't always the ones who represent the broader population. To avoid this bias, we worked closely with a community partner who had deep ties to the local area. They helped us identify participants with a range of experiences and engagement levels—from those actively involved in community organizing to people who were more withdrawn or hesitant to engage.

By making sure we had a diversity of voices, we were able to gain a more nuanced understanding of the challenges and opportunities within the community. What stood out most was how grateful the quieter participants were to have their perspectives considered. They often mentioned that they hadn't been asked for their opinion before, and their appreciation reinforced the importance of reaching beyond the usual voices. This approach not only made our final solution more robust but also strengthened the community's trust in the project. They felt genuinely included rather than tokenized or sidelined.

Ethical Compensation Without Creating Imbalances

Compensation is another area where power dynamics come into play. In many projects, the question of how to appropriately compensate participants for their time and input can be tricky. Offering monetary compensation can sometimes create uncomfortable dynamics, particularly if the community is economically disadvantaged. On the other hand, not compensating people can feel exploitative, as if we're taking from the community without giving back.

In one project for a high-end personal care brand, we found a middle ground that worked well. Instead of offering cash, we gave participants products from the company as a form of compensation. This showed immediate appreciation for their contributions while also creating a positive connection between the community and the brand. The participants were thrilled to receive something tangible for their efforts, and because these were high-quality, desirable products, the gesture felt meaningful. At the same time, it avoided the complications that can arise from direct financial compensation, such as concerns over how the money might be perceived or used.

This method of compensation also reinforced a sense of partnership rather than a transactional exchange. It allowed us to thank participants in a way that felt both thoughtful and equitable, strengthening the bond between the community and the brand without muddying the waters with cash.

Bridging Cultural and Social Gaps

In every project, designers must be sensitive to the fact that they're often entering communities where cultural or social norms differ from their own. It's important to recognize the limitations of your understanding and seek out ways to bridge these gaps. This is particularly true in Equitable Design, where failing to understand or respect the community's values can lead to solutions that miss the mark entirely.

The real estate project mentioned earlier, where we engaged with marginalized communities around housing practices, is a good example. We weren't creating solutions in a vacuum; we were entering a world that had its own complex dynamics. By providing clear explanations and always keeping communication open, we built a relationship with the community that was grounded in mutual respect. But even with all the planning, we remained aware that we were guests in their space and that listening, rather than dictating, was the key to making real progress.

Shared Ownership and Empowerment

Finally, one of the most important ways to manage power dynamics in community engagement is sharing ownership of the project's outcomes. When community members feel like active contributors rather than passive subjects, they are more likely to engage meaningfully. In the banking services project, for instance, we made it clear that the community's feedback would directly shape the final product. This wasn't just lip service—we took their ideas seriously, integrating them into the design and keeping people informed about how their input was being used.

The result was a better solution as well as a deeper sense of partnership between the designers and the community. People felt empowered because they weren't just participants; they were co-creators. And that's the real goal of Equitable Design: to create systems and solutions where the people affected by the outcome have a hand in shaping it.

Addressing ethical considerations and power dynamics is critical to the success of Equitable Design projects. By being transparent, inclusive, and thoughtful about how we engage with communities, we can create solutions that are both more effective and more just. Whether through clear communication, thoughtful compensation, or the inclusion of a broad range of voices, the goal is always to balance the scales and ensure that everyone involved feels respected and empowered. In the end, the most successful projects are those where the community feels not only heard but also truly understood and valued.

Notes

[1] World Intellectual Property Organization (WIPO), 2010. When innovation is child's play. [Online]. Available at: `https://www.wipo.int/en/web/ip-advantage/w/stories/when-innovation-is-child-s-play` [Accessed: 15 January 2025].

[2] Borland, R., n.d. The PlayPump. Objects in Development. [Online]. Available at: `https://objectsindevelopment.net/the-playpump` [Accessed: 15 January 2025].

[3] InformIT, n.d. Case study from creating breakthrough products: the OXO GoodGrips. [Online]. Available at: `https://www.informit.com/articles/article.aspx?p=24132&seqNum=4` [Accessed: 15 January 2025].

[4] Willett, L., 2018. Inclusive design day 11-15: OXO figured it out. [Online]. Available at: `https://fullstackleader.blog/2018/02/12/inclusive-design-day-11-15-oxo-figured-it-out` [Accessed: 15 January 2025].

Implementing Inclusive Solutions and the Business Cases Behind Them

The principles of Equitable Design are not only a moral imperative but a strategic advantage. This chapter explores two critical truths that drive its adoption. First, Equitable Design leads to better products and services that benefit everyone, from safer cities to more intuitive interfaces. Second, contrary to common misconceptions, implementing these principles does not incur additional costs. In fact, by expanding the customer base and addressing underserved markets, Equitable Design often enhances profitability and resilience.

This chapter delves into the economic case for inclusivity, with real-world examples demonstrating how equitable solutions drive growth, improve customer loyalty, and open new revenue streams. We'll explore projects that range from developing scalable products for small businesses to creating more inclusive financial services, illustrating how organizations can align profitability with social responsibility.

Equitable Design doesn't just make products more accessible—it redefines the boundaries of innovation, attracting new customers, minimizing risk, and setting industry benchmarks. By the end of this chapter, you'll see how integrating these principles both transforms lives and strengthens the bottom line, proving that inclusivity and profitability are not mutually exclusive but deeply intertwined.

Business Cases: Four Ways to Make the Investment Count

When I first developed the Equitable Design framework, I genuinely thought the mission alone would be enough to get people on board. After all, who wouldn't want to help make the world more equitable? Who wouldn't want to invest a little to make their products more inclusive and accessible? But to my surprise, that wasn't the reality. Although people were supportive of the cause and agreed on its importance, they struggled to align it with their immediate priorities and short-term business goals.

I quickly realized that I had to meet decision-makers where they were: focused on profitability. Some were upfront, saying, "If we can create equitable solutions and see a financial return, then we're all in. Otherwise, it's going to be a tough sell."

At first, I found this a bit disheartening, even cynical. But I soon recognized the opportunity. I knew that inclusivity and equity lead to better performance and, ultimately, financial gain. If I could demonstrate that, we'd be speaking the same language, and our projects would have a real shot at success.

In this section, I'll outline four models for creating better business through Equitable Design, backed by examples from my work. These strategies have been crucial in turning ideas into tangible results and driving real implementation.

As I'm doing throughout this book, I will purposely connect these business models with the projects discussed earlier in the book, providing an end-to-end perspective that highlights how we worked to enhance the chances of success and implementation. The goal is to illustrate financial growth drivers with projects you are now familiar with. I will also add some examples you may have heard of or that are well-documented outside of my corpus of work.

Product Inclusion as a Growth Driver

One of the key business cases I developed centers on the idea that inclusive products act as powerful growth drivers. If a product is designed so that more people can use it, you naturally expand your addressable market, which leads to more revenue. When these products are built on the foundation of existing infrastructures, the cost to serve additional customers is often lower, which keeps margins healthy and boosts overall profitability.

Take the United States as an example. According to the U.S. Census Bureau, by 2045, the country will become a majority-minority nation, meaning over 50% of the population will be non-white. In that scenario, designing for minority groups essentially means designing for the majority. For companies that want to capture a significant portion of the market in the years to come, this is not just a moral imperative but also a business one.

From Corporations to SMEs: Expanding Reach with Scalable Solutions

In one particular project, we collaborated with a leading global tech company that had traditionally focused on serving large corporations. Over the years, it had built a strong reputation by adapting to major technological changes and maintaining a world-class client base. But as the market evolved, it recognized the need to expand beyond its established niche.

Our key insight during this project was to shift the company's focus towards small and medium-sized enterprises (SMEs). Not only did this align with the goal of offering equal access to advanced tools for smaller businesses, but it also presented a huge growth opportunity. SMEs represent a vast, fragmented market, and because the company's AI and software products were easily scalable, the potential to serve hundreds of thousands of new clients was very real.

Equipping these smaller businesses with the same tools typically reserved for large corporations also had a strategic benefit. Many SMEs regularly interact with larger companies, and IT system compatibility is crucial. By providing its services to smaller players, the company indirectly strengthened its presence among major enterprises as well.

Additionally, because the business model relied on subscription services and leveraged the immense data its clients generated, expanding their user base offered long-term value. The company could boost resilience, create lasting relationships, and increase the overall value of its assets by onboarding new users into the system.

In this case, the project's success hinged on our ability to design the software and services to be easily accessible for non-experts while ensuring scalability to serve millions of users. By keeping the development process universal, the company could seamlessly deploy its solution across a vast range of clients without making it overly complicated or resource-intensive. This approach allowed it to broaden its reach efficiently, expand its customer base, and grow its influence in both small and large markets without a heavy lift in customization or support.

Inclusive Banking: Expanding Reach and Revenue Through Better Design

The challenge of offering financial services to unbanked and underbanked users holds significant potential for revenue growth.

Banks have an opportunity to attract new customers, particularly by redesigning their products and experiences to reach unbanked or underserved populations. Although average revenue per customer varies based on account types and services, the long-term profitability of bringing new clients into the banking system remains clear. For example, a standard retail checking account typically generates around $112 in annual revenue, and interest-bearing accounts can bring in much more—averaging $1,639 per year, largely driven by net interest income. When scaled, the costs of acquiring these new customers through thoughtful redesigns are likely to represent only a fraction of traditional acquisition expenses. Additionally, new customers often unlock further revenue opportunities, such as loans, which not only provide significant financial returns for the bank but also empower clients to move into higher-value banking segments, creating a mutually beneficial growth trajectory.[1]

In the project I mentioned earlier with a large European bank, the primary motivation was to improve the user experience for underbanked and unbanked populations as part of the bank's effort to solidify its positioning as the "bank for everyone." Although the initial focus was on addressing this strategic necessity, after the CEO was publicly challenged on the issue during a local radio interview, we went beyond meeting this goal by demonstrating how inclusivity could also be a profitable opportunity. We knew that with redesign efforts like simplifying interfaces and making services accessible in different languages, the project would both enhance the accessibility of the bank's offerings and lead to increased enrollment of new clients. This new customer base wouldn't just bring in direct revenue; it would also open up cross-selling opportunities for other services, creating a ripple effect across the business.

But beyond the immediate financial return, it's crucial to consider the social and economic impact of providing accessible financial services to communities that are often excluded. When underserved communities gain access to banking services, they can improve their lives in fundamental ways—gaining credit for housing, loans for education, or capital to start small businesses. This isn't just a business opportunity; it's a social one. In this regard, the work echoes the success of microfinance institutions like Grameen Bank, which has shown how low-income populations can drive significant returns when given the chance. The Grameen Bank, founded in 1976 by Muhammad Yunus in Bangladesh, revolutionized microfinance by providing small, collateral-free loans to low-income individuals, particularly women. Its success in fostering entrepreneurship and lifting communities out of poverty earned Yunus and the bank the Nobel Peace Prize in 2006, highlighting the profound social and economic impact of financial inclusion.

At each stage of the project, my team of designers and I quickly identified the business case for inclusivity, and we honed in on the gaps that needed to be addressed. Simple adjustments—such as modifying user interfaces, adding extra features for digital services, and changing the design of physical products like bank cards—created immediate, impactful benefits for both the users and the business. These changes were relatively inexpensive to implement, but they greatly enhanced accessibility and customer engagement, increasing the company's reach while also improving customer satisfaction.

In many of the projects I led as a consultant or a teacher, my team and I were able to identify both the business opportunity and the gaps that needed to be filled to make the product more inclusive. Whether through small adjustments to an interface, the addition of new digital features, or tweaks in the packaging's color and language, these changes were relatively simple to implement but had a big impact on both the users and the company.

Strengthening Customer Loyalty Through Inclusive Branding

One key way to drive growth with Equitable Design is by increasing customer retention. For brands that offer commoditized products like shampoo or shower gel, retaining customers can be challenging, as consumers tend to switch between brands unless there's a strong value proposition. This is especially true in the mid-range market, where brand loyalty is harder to secure than it is for premium products that often have an emotional appeal or luxury factor.

The importance of inclusivity in building loyalty is backed by compelling research. A global study by the Unstereotype Alliance, conducted with Oxford University's Saïd Business School, found that inclusive advertising led to a 15% increase in customer loyalty and made consumers 62% more likely to choose these brands as their first choice. Beyond loyalty, these campaigns drove a 3.5% increase in short-term sales and a remarkable 16% lift in long-term sales. Similarly, Kantar's Brand Inclusion Index highlighted that 75% of consumers consider a brand's reputation for diversity and inclusion when making purchase decisions. Brands with strong inclusivity practices saw 33% higher consumer consideration, underlining the direct link between inclusivity and customer retention. Furthermore, surveys reveal that 57% of consumers are more loyal to brands that address social inequities, and 50% are more likely to recommend brands that reflect diversity and representation. These findings emphasize that aligning with consumer values not only fosters trust but also drives tangible business results.[2-4]

Linking this back to the shampoo example I mentioned earlier, our goal was to shift consumer behavior and encourage deeper loyalty by revamping the brand's packaging, messaging, and overall identity to make it more inclusive. We focused on gender and ethnicity, creating a brand image that resonated with a wider audience. The key here was to make the brand stand out on store shelves—not just as a product, but as a statement.

One of the challenges we encountered was the traditional separation of products by gender. Shampoo aisles are typically divided into sections for men and women, but our goal was to design a product that defied those conventions. This led us to create gender-neutral packaging combining colors and textures that weren't typically associated with one gender. The messaging on the packaging also directly addressed inclusivity, making a bold statement about gender and ethnic equality.

The impact we were aiming for went beyond just making the product visually appealing. We wanted it to spark conversations among consumers—whether in the store or later, at home. Imagine a shopper picking up this shampoo, seeing its bold design and inclusive message, and then discussing it with their friends or family. That kind of emotional connection and social endorsement can turn a casual shopper into a brand ambassador. In marketing, once a customer becomes an ambassador, their loyalty is cemented, and they start attracting new customers as well.

This approach doesn't just create one-time sales; it fosters long-term loyalty. Customers feel a deeper connection to the brand because it aligns with their personal values around inclusivity and equality. For us, this project was a prime example of how Equitable Design can go hand in hand with business growth, turning consumers into advocates and increasing both retention and brand visibility.

It seems we're destined to stay in the world of shampoo and personal care products—but this time, let's turn our attention to a campaign that has become a case study in inclusive branding: Dove's "Campaign for Real Beauty" (see Figure 7.1). Launched in 2004, this groundbreaking initiative set out to challenge traditional beauty standards by showcasing women of diverse shapes, sizes, and ethnicities. The campaign didn't just resonate; it redefined how beauty was portrayed in advertising, earning widespread praise for its authenticity and inclusivity. Importantly, it also proved a powerful business case: customer loyalty surged as women connected with the brand's empowering message, and Dove's market share grew significantly as a result. Evidence of this loyalty can be found in the numbers—by 2006, two-thirds of Dove's sales were generated by people who purchased more than one Dove product, doubling the figure from 2003, before the campaign launched.

But the road wasn't entirely smooth. Critics questioned the sincerity of Dove's commitment to inclusivity, pointing out that its parent company, Unilever, also owned brands that perpetuated conventional beauty ideals. Additionally, some saw the campaign as more of a savvy marketing move than a genuine effort to redefine beauty standards. These controversies, however, didn't overshadow the campaign's broader success. The "Campaign for Real Beauty" demonstrated how inclusive branding can build lasting loyalty and elevate a brand's standing—even in the face of criticism—proving that aligning with consumer values is not just good ethics but also good business.

Expanding Social Media Reach with Inclusive Features

In the world of social media, it's clear that when we aren't paying for a product, we often *are* the product. This model, which involves selling users' data and attention to advertisers, has fueled the success of platforms like Facebook,

Google, and other social networks. With this understanding, the work we did for companies in this industry also had a direct return on investment—by making the platform more inclusive, we were able to attract more users and drive higher engagement.

Figure 7.1: Dove's "Campaign for Real Beauty" in the London Subway in 2004

An example from my own experience, which I mentioned earlier, comes from the work we did with a team of students and a major social media platform to improve the experience for Black business owners. By designing features that thoughtfully addressed their needs—such as customizable tools for promoting their businesses, specific buttons or reactions tailored to their audience, and a more transparent, user-friendly algorithm for recommendations—we helped attract new users to the platform. These business owners had previously avoided using the platform because they didn't realize its potential and didn't feel included. With these improvements, not only did the business owners join the platform, but—by sharing their presence on social media—they brought their followers along, significantly expanding the platform's reach.

Beyond just increasing user numbers, this initiative also boosted engagement. Communities that previously felt underrepresented or underserved now had more connections with the contents of the platform and a reason to spend more time on it, engage more meaningfully with it, and feel a stronger sense of belonging. These relatively simple design adjustments created a ripple effect, increasing user activity and interaction, which in turn provided the platform with richer data for its advertisers.

This case shows that Inclusive Design doesn't just create goodwill; it can also have immediate, measurable business benefits. When we align the design of a product with the business model, especially in the data-driven world of social media, we can see real returns in terms of botj user engagement and profitability.

Expanding the Pool of Partners, Talent, and Stakeholders

Another significant business case for embracing Equitable Design and Inclusive Design is the way it allows companies to expand their network of partners, talent, and stakeholders. Companies don't operate in isolation; they're much more dependent on external collaborators and relationships than they might realize. The pandemic was a stark reminder of this interdependence. As supply chains broke down and familiar systems became unreliable, it highlighted how interconnected our world truly is—and how much we lack resilience when that network is disrupted.

By broadening the range of partners and collaborators, companies can create more diverse and robust networks that help them not only grow but also weather difficult times. When organizations actively work to include a wider array of perspectives and backgrounds, they access new ideas, innovations, and resources that can significantly strengthen their position in the market. This diversity of thought and expertise isn't just about surviving in challenging circumstances; it's a pathway to thriving in a rapidly changing world.

Leveling the Playing Field for Startup Success

In a project with a major startup sourcing platform, we aimed to tackle an issue that's often overlooked: the disparities in access and opportunity for startups. The platform's main function is to connect large corporations and public institutions with startups that have innovative solutions to their challenges. Although the basic model works, we noticed a significant imbalance in the startup landscape. Startups founded by people from privileged backgrounds tend to have an easier time securing funding, getting meetings, and being introduced to potential business partners. They often possess strong networks and connections, which can open doors before the startup even begins proving itself.

On the other hand, startups founded by less privileged entrepreneurs face higher hurdles. Without the same level of access or connections, they often struggle to get noticed, even if they have a compelling product or solution. The problem is that when opportunity is skewed in this way, companies and institutions looking for innovative solutions may miss out on some of the most creative or impactful ideas simply because the founders lacked the "right" background or network.

Recognizing these disparities, we collaborated with the platform's team to establish a more inclusive process for discovering and evaluating startups. The aim was to make it easier for talented entrepreneurs from all backgrounds to get their ideas in front of the right people.

At the same time, the institutions using the platform are typically looking for unique solutions—ones they haven't found through traditional channels. Their expectations are high, hoping this scouting platform will unearth promising startups they otherwise wouldn't have encountered.

In this context, our goal was to help the platform create an inclusive process that would reach out to, evaluate, and support startups without bias toward their background or origin. The result was a more equitable system that provided equal access to opportunities for all startups and a broader, more diverse pool of potential solutions for the clients.

The inclusive approach delivered a clear return on investment (ROI) for all parties involved. For startups, the revamped process opened doors to funding and partnership opportunities that had previously been out of reach, enabling them to focus more on refining their solutions rather than struggling to gain attention. For the sourcing platform, an inclusive process translated into higher engagement across diverse founders, which broadened the pool of talent and ideas available. This variety also made the platform more attractive to large corporations and institutions looking for innovative solutions, as it increased the likelihood of finding unique answers to their challenges.

On the client side, companies and public institutions gained a competitive advantage by tapping into a previously overlooked reservoir of diverse perspectives and solutions. This often led to the discovery of new approaches that traditional networks missed. Ultimately, the platform's ability to offer a wider, more representative portfolio of startups created a win-win situation where both the quantity and quality of potential solutions increased. The investment in an unbiased and equitable process was justified by the tangible benefits of stronger partnerships, innovative problem-solving, and more meaningful engagements across the ecosystem.

Unlocking Workforce Potential Through Inclusive Job Design

This business case appeared clearly to me when I consulted for the wine producer in Napa Valley. We conducted a detailed assessment of its workforce and discovered a significant turnover rate in certain job segments. Delving into the causes, we found that these roles often had challenging working conditions, such as exposure to loud noise or physically demanding tasks, which made them less appealing to employees. The high churn cost the company time and resources in recruitment and training.

However, by shifting the perspective and focusing on how certain conditions might align well with the unique skills or needs of some individuals, we found a new path forward. For example, deaf workers could thrive in roles that involved high noise levels, as they wouldn't be affected by the auditory strain that impacted other employees. Similarly, individuals with ADHD often performed exceptionally well in physically active jobs that provided constant movement and stimulation. Additionally, people with heightened senses of taste or smell, such as those who were visually impaired, could bring a unique advantage to roles like wine tasting and blending, where sensory precision is essential.

By recognizing these so-called "extra-abilities" rather than focusing on disabilities, we not only identified ways to optimize productivity but also saw an opportunity to improve job satisfaction and retention. The approach turned a perceived problem into a strategic advantage, matching job requirements with the strengths of a diverse workforce. The concept of employing people with these unique skills as specialists turned out to be a more efficient use of talent, as it aligned the tasks' demands with the workers' strengths, leading to better performance and longer employee tenure.

This experience highlighted how inclusive and equitable job design can present immediate solutions to persistent problems. Rather than trying to fit employees into traditional roles, rethinking job structures to accommodate different abilities yielded a quick return on investment by reducing turnover, increasing productivity, and fostering a more inclusive and loyal workforce.

Rethinking Recruitment: Expanding Talent Pools for Growth

In 2023, we partnered with one of the world's largest insurance companies, a well-established brand that had built a highly successful business over the years. Despite its solid reputation, the company faced a significant challenge: attracting and retaining new talent. Although it had made substantial investments in promotional campaigns and hired more HR experts to devise ambitious internal strategies, the company struggled to make meaningful progress. The initiatives led to only marginal improvements, indicating a need for a different approach.

When we came on board, our mission was to find alternative ways for the company to attract new talent and retain those who had recently joined. Instead of trying to compete head-on with more appealing industries, we suggested focusing on untapped talent pools—communities with less exposure to traditional insurance career paths but with the same potential for skill and contribution. This strategy included rethinking the company's branding, hiring process, and communication efforts to reach a wider, more diverse audience.

Our main challenge was to make the insurance industry appealing to young professionals who might never have considered working in such a traditional field. The issue was largely cultural: we needed to connect with these talents by speaking their language, understanding their motivations, and creating a sense of belonging. The goal was to generate a positive ripple effect where once a few candidates joined, they could share their experiences and draw in more peers from their networks.

For the insurance company, the business case was clear. It needed fresh talent to sustain growth and operations, and reducing the cost of recruiting and retaining staff was a priority. By opening doors to an overlooked pool of skilled professionals, the company could address both issues while also benefiting communities by offering stable, well-paying jobs with clear career growth paths. This approach diversified the company's workforce and strengthened its long-term talent pipeline and brand reputation.

Another compelling example of expanding talent pools through inclusive practices comes from IBM's partnership with Historically Black Colleges and Universities (HBCUs) through the IBM-HBCU Quantum Center. Launched in 2020, this initiative aimed to develop a more diverse workforce in the burgeoning field of quantum computing, where underrepresentation of marginalized communities remains a significant challenge. By collaborating with over 20 HBCUs, IBM provided students and faculty with access to quantum computing resources, training, and research opportunities.

For IBM, the business case was clear: ensuring a steady influx of skilled, diverse talent in an industry expected to grow into a $125 billion market by 2030. This partnership allowed IBM to secure its leadership in quantum computing by broadening its talent pipeline while aligning itself with global ESG (environmental, social, and governance) goals. The collaboration also enhanced IBM's reputation as an innovator committed to equity, which bolstered its relationships with key stakeholders, including governments and public institutions that are critical clients for quantum technologies.

However, the initiative faced challenges, such as disparities in resource utilization among participating HBCUs with varying levels of STEM (Science, Technology, Engineering and Math) infrastructure. Additionally, critics have called for greater transparency regarding tangible outcomes, such as the number of students entering quantum computing careers or securing roles at IBM. These hurdles highlight the importance of building comprehensive support systems to ensure that such partnerships create sustained, meaningful impact.

Despite these challenges, the IBM-HBCU Quantum Center demonstrates how inclusivity can serve both societal and business goals. By investing in overlooked communities, IBM not only strengthens its workforce but also sets a precedent for how equitable practices can drive innovation and long-term growth.[5-7]

Redefining Industry Standards to Stand Out

Inclusive and Equitable Design doesn't just make products and services accessible to more people—it can also help companies set new benchmarks in their industry.

When businesses invest in this kind of design, they can disrupt the status quo, reshaping how customers make purchasing decisions, whether consciously or unconsciously. Companies that successfully implement inclusive practices can create a unique selling point that sets them apart in crowded markets, giving them an edge in visibility, sales, and even negotiation power.

By pioneering more equitable standards, companies don't just follow market trends—they redefine them, influencing competitors and raising expectations across the industry. This kind of leadership benefits customers by offering more accessible and thoughtful products and also positions the company as a forward-thinking brand known for driving progress and innovation.

Again, here, I will circle back to projects I introduced in earlier chapters to show how they translated into a beneficial business case for the partner companies.

Setting the Benchmark for Seamless Authentication

When we started collaborating with a leading communications company, the project initially focused on its authentication service. The goal was to explore how making cybersecurity more accessible could also present a profitable business opportunity. In an increasingly digital world, ensuring that everyone has secure and convenient access is vital—but it was not immediately clear how this would translate into a compelling business case.

The problem we tackled was making the two-step authentication process more inclusive for people with visual impairments. For many users with these challenges, voiceover assistance was the main way to navigate security prompts. This, however, posed a significant risk, as the voiceover would read sensitive information aloud, potentially exposing passwords to anyone nearby.

We explored various solutions, such as using haptic feedback for a more tactile experience and creating simplified interfaces that would be easier to navigate. But the breakthrough came when we started considering the use of geolocation. By positioning a user's devices—such as a laptop, phone, tablet, or smartwatch—within a certain proximity to each other, the system could determine that the login attempt was legitimate. If the devices were near each other, the app would seamlessly authenticate the user without needing a password.

The most exciting realization was that this solution was not just beneficial for visually impaired users—it had universal potential. The convenience of geolocation-based authentication appealed to all users, as it eliminated the need to juggle multiple interfaces, enter passwords, or click through several windows. Testing the prototypes confirmed that users appreciated the speed and seamless nature of the new process, proving it could work for a broad audience.

What began as an effort to enhance digital security for a specific group turned into a universally adaptable solution that addressed inclusivity and improved the user experience for everyone.

As we presented the idea to our client, we emphasized some key metrics that illustrated the potential impact of the solution:

- **The authentication process would involve:**
- 0 vocal interfacing
- 0 switching between apps or screens
- **Saving:**
- 1 double-tap gesture
- 9 swiping gestures
- 27 seconds saved per authentication

When we shared these figures, there was a moment of silence in the room. We quickly realized that when scaled, the time savings could be truly significant. The client's service was used by one of the largest tech companies in the world, employing hundreds of thousands of people globally, many of whom had among the highest salaries in the industry. By saving 27 seconds per day for these employees, we calculated that it would add up to roughly half a million hours of productive time saved annually—translating into millions of dollars in cost savings.

This approach set a new industry benchmark: authentication should be seamless, efficient, and designed to maximize users' productive time. The business case practically wrote itself, proving to be a compelling argument not only for our client but also for its customers, who stood to benefit greatly from a streamlined and time-saving process.

Our inclusivity-driven approach in this project underscores a larger point: designing for inclusivity is not just about catering to marginalized populations—it often leads to solutions that benefit everyone. The *curb-cut effect*, a term born from the addition of sidewalk ramps originally intended to improve accessibility for people in wheelchairs, exemplifies this beautifully. Curb cuts transformed mobility for wheelchair users, but they also made life easier for parents pushing strollers, delivery workers with carts, kids on skateboards, and countless others. What began as a targeted solution became a universal improvement.

Similarly, the geolocation-based authentication system we developed initially aimed to address the unique challenges faced by visually impaired users. Yet its benefits extended far beyond that group. The simplified, seamless process improved security and convenience for every user—whether navigating tight schedules at work, multitasking with limited attention, or simply appreciating a more intuitive experience. By addressing a specific need with inclusivity in mind, we created a solution that raised the bar for all users.

From a business perspective, this universal appeal is where inclusivity becomes a game-changer. Companies that invest in Equitable Design unlock access to broader customer bases and drive higher engagement, loyalty, and satisfaction.

These inclusive innovations often yield exponential returns, as their value proposition resonates with far more people than initially anticipated. In this case, the geolocation authentication not only bolstered security and efficiency but also showcased how inclusivity-driven design can evolve into a universally desirable feature, creating a significant competitive advantage.

Making Accessibility Mainstream: Panasonic's Industry-Leading Approach

The examples discussed earlier stem from projects I have directly participated in, but Panasonic's universal design initiative offers a notable industry case study of how inclusivity can serve as both a social responsibility and a business advantage. This example underscores how a major corporation has successfully leveraged accessibility principles to drive innovation, expand its market reach, and enhance its reputation in a competitive global landscape.

Building on the theme of making services more accessible while driving business growth, Panasonic's universal design initiative offers a compelling parallel to the concept of seamless authentication. Much as inclusive authentication solutions simplify processes for all users, Panasonic has embraced universal design to create products that are intuitive and user-friendly for everyone, regardless of age or ability. This approach has not only demonstrated their commitment to accessibility but also delivered substantial business benefits.[8]

Panasonic applies universal design principles across a range of products, from home appliances like microwaves and washing machines to personal care items and consumer electronics. For instance, the company's washing machines feature simple control panels with large buttons and clear symbols, making them easier for older adults and individuals with limited dexterity to use. Similarly, Panasonic's televisions incorporate voice guidance for users with visual impairments and intuitive remote controls that enhance usability for all.

The business impact of this approach has been significant. By focusing on universal accessibility, Panasonic has expanded its market reach, particularly among aging populations in Japan and other developed countries. Japan, for example, has one of the world's most rapidly aging populations, with nearly 30% of citizens aged 65 or older.[9] By designing products that cater to this demographic, Panasonic has not only addressed a social need but also positioned itself as a market leader in a lucrative and growing segment. The global market for assistive devices for elderly and disabled individuals was valued at over $20 billion in 2020 and is projected to grow significantly, a trend Panasonic is well-poised to capitalize on.[10]

Moreover, these universally designed products have garnered widespread acclaim, enhancing the brand's reputation. Kantar Research indicates that brands with inclusive advertising experience a 15% increase in customer loyalty and a 62% higher likelihood of being a consumer's first choice. For Panasonic, this has translated into stronger customer relationships and increased market share in key regions.

Although Panasonic's universal design initiative has been a game changer in many respects, it is not without its challenges. One limitation is the balancing act between creating universally accessible products and maintaining competitive pricing. Features like voice guidance and specialized ergonomic designs often come with higher production costs, which can make these products less appealing in price-sensitive markets. These features add value, but some consumers may opt for cheaper alternatives that meet their basic needs, potentially limiting the widespread adoption of inclusive products.

Additionally, universal design, although inclusive, can sometimes fail to address the specific needs of niche user groups. For example, products designed with general accessibility in mind may not provide sufficient customization for individuals with highly specialized requirements, such as advanced hearing aids or adaptive communication devices. This underscores the importance of striking a balance between universal design and targeted solutions.

Panasonic's universal design journey demonstrates that inclusivity is not merely an ethical imperative but also a profitable business strategy. By creating products that prioritize ease of use and accessibility, the company has expanded its customer base, improved brand loyalty, and tapped into growing markets. However, as with any inclusive initiative, success requires continuous refinement and an understanding of where universal solutions may fall short. This case, much like the example of seamless authentication, highlights that Inclusive Design has the power to make life better for everyone—and generate tangible returns for businesses that embrace it.

Beyond Syntax: Redefining Writing Tools for Inclusive Impact

When we partnered with a company that offers digital writing software focused on correcting spelling, the opportunity arose to push the boundaries of what their service could offer. Our aim was to go beyond simple syntax and grammar correction by including edits and suggestions that promote inclusive and respectful language. We envisioned a tool that wouldn't just correct spelling but would also help users communicate in ways that foster inclusivity and understanding in both professional and personal settings.

Implementing this idea meant investing in training the company's language model to understand and flag both grammatical issues and language that could be considered non-inclusive or potentially offensive. It wasn't just about adding a new feature; it required a deeper understanding of the contexts in which certain words and phrases were used, ensuring that any suggestions made would genuinely enhance inclusiveness without being overbearing. This was a significant leap for our partner, as they felt it took them outside the realm of their original product mandate and user expectations. The company worried about whether users would see this as an overreach or as a valuable addition.

However, to drive meaningful change, sometimes you have to take bold steps. By choosing to integrate these inclusivity features, the company had the chance to set a new standard in the industry. The message was clear: language tools should do more than just fix typos; they should also support users in creating a more inclusive environment. The tool could become a trusted partner for users, helping them navigate not just written communication but the social and emotional dynamics that come with language.

The business case became compelling when we looked at the broader impact. For organizations, communication plays a crucial role in workplace culture and productivity. Companies that foster inclusive communication are better positioned to attract diverse talent, retain employees, and maintain a positive brand reputation. By equipping users with a tool that promotes inclusivity, the software company could help businesses cultivate more harmonious workspaces, which could, in turn, drive better employee engagement and higher productivity.

From a financial perspective, offering these advanced language features created an opportunity to position the software as a premium product. This differentiation allowed the company to potentially command higher subscription fees while tapping into a growing demand for tools that support diversity, equity, and inclusion (DEI) initiatives. The impact went beyond just the individual user; organizations could see tangible benefits in improved team collaboration, fewer workplace conflicts, and an overall boost in morale. As companies reported better communication outcomes, word-of-mouth and organic growth would naturally follow.

The long-term vision was to shift user expectations across the industry. If this tool could set a precedent, it wouldn't be long before other writing assistants would be expected to offer similar inclusive features. Thus, by leading the way, the company wouldn't just be keeping up with trends but actively shaping them. This initiative showed that with thoughtful investment in Equitable Design, it's possible to create solutions that are not only socially responsible but also highly profitable.

Breaking the Mold: A New Approach to Streamlined Content Selection

Streaming services are a battleground, with companies constantly striving to outdo each other in creativity, original content, and user engagement. As Reed Hastings, Netflix's co-founder, said in 2017, "You get a show or a movie you're really dying to watch, and you end up staying up late at night, so we actually compete with sleep." With companies like Netflix, Hulu, and Amazon investing heavily in proprietary content and aggressive marketing, the competition is fierce.

However, a less explored aspect of competition is the user interface. Although a lot of effort has gone into optimizing the streaming experience, the interfaces are typically designed for people with good vision and familiarity with navigating dense information. This often leaves users with low vision struggling, as these interfaces can be cluttered and overwhelming. Those who rely on audio description or other accessibility tools often face significant cognitive fatigue just trying to navigate the menus.

Our project set out to change this by reimagining the interface from the ground up. We didn't want to just tweak the existing design; we wanted to rethink the entire experience, especially for users with visual impairments. Starting with a clean slate, we focused on creating a streamlined decision-making process, using conversational tools and closed-ended questions to help narrow down the user's preferences quickly. This approach allowed us to guide users toward a selection with just five questions, balancing simplicity and personalized recommendations.

The real breakthrough came when we realized that this solution wasn't just beneficial for users with low vision—it had universal appeal. During testing, we noticed the benefits extended to anyone trying to multitask. Whether it was a driver picking a show while heading home, parents choosing a movie while tidying up after putting the kids to bed, or children starting their favorite cartoons without needing help from an adult, our solution fit seamlessly into everyday life.

By simplifying the decision-making process and removing the cognitive burden, we set a new standard for how people interact with streaming services. The ability to choose a program without needing to sit in front of a screen and, with minimal effort, became a game-changer. As this approach becomes more widely adopted, our client will be positioned as a leader in the entertainment space—pioneering a user-centered innovation that breaks the mold in a crowded industry.

Transforming a Commodity: How Inclusive Design Can Revitalize a Notebook Brand

Coming back to previously mentioned cases, when we partnered with a leading notebook company to explore ways of designing products for people with neurodivergence, we encountered a significant challenge: how to innovate in a commodity market. In the world of notebooks, products are highly price sensitive, and consumer decisions are often driven by cost rather than brand loyalty. This creates an environment where companies hesitate to invest in new features, as the returns may not seem immediately apparent.

Our approach was to go beyond simply redesigning the notebook itself. We focused on creating a truly inclusive product line that could stand out in a crowded market. This involved rethinking the notebook's design elements—from the layout of the lines on the paper to the tactile experience of the cover. We also proposed a companion app to bridge the gap between traditional paper and digital tools, offering features like converting handwritten notes into text or providing prompts for neurodivergent users. Initially, the client was skeptical, questioning whether the development costs would be justifiable in such a competitive and low-margin market.

The turning point came when we explained the scale of the opportunity. People with neurodivergence—such as ADHD, autism, or dyslexia—represent about 15 to 20% of the global population. If our partner could be the first to address their needs with even modest improvements, we could capture a significant market share. The inclusive design would not only appeal to neurodivergent users but could also resonate with anyone who values diversity, potentially shifting market perception and positioning the brand as an inclusive leader. This strategy would naturally differentiate the company from competitors, most of whom lacked any features or narrative around inclusivity.

Two additional factors supported the business case. First, many of the proposed changes, such as modifying the app's software and slightly altering the physical design, incurred only initial development costs. Once those costs were covered, the marginal cost of producing each additional notebook was negligible. This meant that as sales volume increased, the cost of implementing inclusive features would decrease, driving higher profit margins over time.

Second, we looked beyond the traditional retail distribution model, which is heavily seasonal due to back-to-school sales. Instead of relying solely on big-box stores, we explored partnerships with healthcare professionals and educational institutions that work with neurodivergent children. Occupational therapists, psychologists, and special education teachers often recommend tools to aid learning, and they were eager to endorse products designed with their patients' needs in mind. This opened up new year-round distribution channels where price sensitivity was less of a concern, as the product would be seen not just as a school supply but as a valuable educational aid.

By positioning the notebooks as tools for learning and cognitive support, we also enhanced the product's perceived value. It could be sold at a higher price point in these specialized markets, where parents and educators were willing to invest in high-quality resources. The year-round sales cycle also smoothed out production demands, reducing the reliance on a single, high-stakes sales season and providing more consistent revenue streams.

Ultimately, the Inclusive Design approach provided a multifaceted business opportunity. It allowed the company to break into underserved markets, tap into new distribution channels, and elevate the brand by setting a new industry standard for inclusivity. This wasn't just about social responsibility; it was a strategic move that delivered tangible economic benefits, proving that embracing diversity can directly impact the bottom line.

Minimizing Reputational Risk with Inclusive Design Strategies

The final business case for Inclusive Design revolves around risk management. Companies are increasingly being called out for non-inclusive practices, and the impact of these controversies has only intensified with the rise of social media. Backlashes can spread rapidly, often reaching a large audience, and public denunciations can quickly escalate, damaging a brand's reputation. Once a company is labeled as non-inclusive, it faces the difficult and costly task of rebuilding its image and regaining public trust, along with any financial losses tied to the negative publicity.

Empowering Consumers Through Ethical Transparency

I had the opportunity to work with an early-stage startup in San Francisco, which aimed to empower consumers to make more informed decisions about the companies they support. The concept was straightforward yet compelling: use AI models to scrape publicly available information about companies and then score them against ethical indexes. The intention wasn't to rank companies per se but to provide a clearer picture of what companies stand for, helping consumers understand what ideologies or community engagements their purchases indirectly support.

For example, the startup would compile data from press releases, news articles, donations, and even public statements from company leaders on various ethical issues, such as abortion rights, gun control, LGBTQ+ rights, and immigration policies. The AI would then score companies on a scale for each issue based on how frequently they were associated with pro- or anti-positions. This way, a company could be positioned as strongly pro-abortion rights, neutral, or anti-abortion rights based on the collected data.

The goal was to enable consumers to make informed choices, supporting companies whose values align with theirs. Although the startup is still in its early stages, I believe this type of ethical transparency will become standard practice in the future, pushing companies to adopt more responsible and inclusive business models.

Scaling up this type of service demonstrates how significantly the decisions made by large organizations can impact their reputations and customer loyalty. Companies that invest in Inclusive Design and practices not only protect themselves from potential backlash but also foster deeper connections with their users. For instance, a company committed to inclusivity might avoid a PR crisis surrounding controversial political donations simply by ensuring that its donations align with widely accepted social values.

Consumers today increasingly demand transparency, and they want to support brands that reflect their values. Companies that prioritize ethical and inclusive practices will naturally attract these conscientious consumers, whereas those that neglect these aspects may struggle to maintain customer loyalty. In this way, investing in ethical transparency and Inclusive Design isn't just about doing the right thing; it's a strategic move to minimize reputational risks and build a loyal customer base.

The Risks of Overlooking Data Privacy in Sensitive Markets

We worked with a data company that had invested heavily in cybersecurity, focusing on FemTech—an industry that provides technology-based solutions for women's health, such as period- and fertility-tracking apps. Our project was rooted in the aftermath of a significant data privacy incident in the FemTech space that highlighted the dangers of not prioritizing data protection and inclusive practices.

A well-known case involved the popular period-tracking app Flo, which came under fire after it was discovered that from 2016 to 2019, the app had been sharing sensitive user health data with third-party companies like Facebook and Google. This data included intimate details about users' menstrual cycles and fertility information despite the company's promise to keep such data private. In January 2021, the Federal Trade Commission (FTC) filed a complaint against Flo, accusing it of misleading its users and failing to place any restrictions on how third parties could use the data. Although Flo settled the case without admitting wrongdoing, the scandal caused severe damage to the app's reputation.

To regain trust, Flo introduced an "anonymous mode," allowing users to access the app without providing personally identifiable information. But the damage had been done, and the case served as a cautionary tale for other companies in the FemTech industry. With reproductive health becoming even more politically sensitive in the United States after the overturning of Roe v. Wade, concerns about privacy and data security in FemTech were thrust into the spotlight.

Our involvement with the data company was an attempt to help rebuild trust after the fact, and the Flo scandal clearly demonstrated the risks of neglecting data protection in sensitive markets. It wasn't just about fixing issues after the backlash had hit but about understanding the deeper ethical responsibility companies have when handling intimate data. Our aim was to develop stronger data security measures and clearer communication strategies to ensure that companies could safeguard user information and be transparent about data usage.

We saw this project as a perfect example of how companies must prioritize Inclusive Design and data ethics to minimize reputational risk. In the FemTech industry, where the nature of the data is inherently private, overlooking these aspects could not only lead to regulatory consequences but also deeply damage brand reputation. By addressing these concerns head-on, companies can avoid costly mistakes and foster a deeper sense of trust with their users, setting a new standard in the industry for privacy and ethical data practices.

Addressing AI Bias to Prevent Reputational Damage

In 2015, Google Photos faced a significant controversy when its AI image recognition software mistakenly labeled a Black person as a "gorilla," an incident that highlighted the racial bias inherent in AI systems. The error was brought to light by Jacky Alcine, a programmer who discovered that the system had categorized photos of himself and a Black friend under this offensive label. The backlash was immediate and widespread, prompting Google to issue an apology and disable the "gorilla" label entirely. However, this response was criticized as more of a temporary workaround than a comprehensive solution to the underlying issues of biased data sets and algorithms.

Google's initial response included promises to improve the accuracy of its image recognition technology, particularly regarding the identification of dark-skinned individuals. Google executive Yonatan Zunger acknowledged the severity of the mistake, stating it was "high on my list of bugs you 'never' want to see happen." He mentioned that the company was working on longer-term fixes involving better recognition of dark-skinned faces and careful consideration of language used in photo tagging. Despite these assurances, many critics felt that Google did not adequately address the systemic biases within its algorithms. Jacky Alcine himself expressed concerns about the types of images and data used to train Google's AI, questioning whether the company would genuinely improve diversity in its workforce to prevent similar incidents in the future.

Years later, reports indicated that Google opted for a workaround by removing the ability for its AI to recognize certain animals, including gorillas, altogether rather than developing a more permanent solution to address the bias problem. This decision was seen as an avoidance of tackling the root causes of racial bias in AI systems. Although Google took steps to prevent further mislabeling incidents, critics argued that simply censoring specific labels did not resolve the broader issues of representation and fairness in AI technology.[11–13]

This case became a classic example in discussions about AI ethics and the importance of Inclusive Design, demonstrating how algorithmic biases can lead to deeply offensive and problematic outcomes. The controversy damaged Google's reputation and raised awareness about the ethical challenges associated with AI, sparking industry-wide discussions on fairness, accountability, and transparency in machine learning.

Years later, while working with a tech company in this industry, we focused on a different aspect of inclusiveness: making phone software more accessible. As we explored the various features, we noticed that accessibility settings were hidden deep in the menu, requiring users to complete the standard setup before they could adjust any accessibility settings. This design flaw treated accessibility as an afterthought rather than a core component of the user experience.

Our suggestion was straightforward but impactful: make accessibility settings available by default during the phone's initial setup. Users would have the option to choose between a standard mode or accessibility mode as the very first step when starting up the phone. This simple change would not only improve the inclusiveness of the product but also signal a commitment to accessibility from the start, helping to reinforce the company's reputation as a leader in Inclusive Design.

By prioritizing inclusiveness in the design process, companies can avoid the pitfalls that lead to reputational damage and initiate a cultural shift among their designers and employees. This proactive approach creates an environment where inclusiveness becomes integral to the development process, reducing the likelihood of future incidents and demonstrating a genuine commitment to diversity and accessibility.

Steps to Implementation

After identifying an inclusive and equitable solution, the real work begins. Moving from concept to execution requires a detailed strategy to ensure the solution addresses the identified problems, aligns with business goals, and remains competitive in the market. The first step is to test the solution rigorously. This involves evaluating its effectiveness in solving the problem, making adjustments based on feedback, and assessing its overall impact on the user experience. Testing ensures that any potential barriers are identified and resolved before the solution is rolled out on a larger scale.

A thorough testing phase is more than just validating technical functionality. It involves engaging diverse user groups in the process to ensure that the solution meets their needs and expectations. Inclusive testing means going beyond the standard user personas and including individuals who represent a broad range of abilities, cultural backgrounds, and perspectives. This approach can help uncover usability issues that may not be immediately apparent but could significantly impact some users' experiences.

Defining a Testing Strategy

The first step in bringing an inclusive solution to life is establishing a solid testing strategy. Many people think of prototyping as building something that closely resembles the final product, but good prototyping is much more intentional. Each prototype should be tailored to test specific aspects of the design rather than aiming for a perfect end product right away. Think of it as crafting distinct tools for different tasks—each prototype should be designed to answer a particular question or test a certain hypothesis.

When we worked on redesigning the interface of a software suite, the need for multiple prototypes quickly became clear. We had a variety of questions:

How do users respond to new features? Can they easily find the buttons and navigate the interface? Are the interactions intuitive, or do users get lost in the process? On top of that, we needed to assess technical elements, such as the speed and accuracy of the features.

These different questions required different types of prototypes. For example, when testing new features, we focused on simplified versions of the interface that highlighted specific functionalities rather than showing the complete design. When evaluating the user experience, we created interactive mockups that mimicked the actual user flow. This allowed us to observe how people navigated the system, where they hesitated, and what they found frustrating. Technical tests often involved stripped-down prototypes that ran in controlled environments to measure processing times and system responses.

But it didn't stop there. Our primary audience was people with neurodivergence, which encompasses a broad range of needs and preferences. This meant creating different prototypes for different types of neurodivergent users—what worked for someone with ADHD might not suit someone on the autism spectrum. And because we wanted the solution to be universal, we brought in additional groups for testing: older adults, people who weren't tech-savvy, and users with visual impairments. We had to ensure that our inclusive design for one group didn't inadvertently exclude others.

A memorable moment during this phase came when we were testing the interface with an older adult who wasn't very familiar with technology. What seemed like a straightforward task—finding a button—turned into a 15-minute struggle. This encounter highlighted the need for clearer visual cues and larger touch targets, which we hadn't initially considered a priority. The prototype revealed an important gap, allowing us to make a simple change that greatly improved accessibility for a broader audience.

This multilayered approach illustrates how defining a testing strategy involves much more than just building a single prototype. It's about crafting different versions to address each aspect that needs to be evaluated and each user group that will interact with the solution. The idea isn't to overwhelm yourself with endless iterations but to start with a few key hypotheses and develop targeted prototypes that can easily be tweaked. You're aiming for maximum learning with minimal effort, which often means keeping things scrappy.

In Chapter 6, "Designing for, with, and by Communities," I emphasized the importance of having an open mind during prototyping. This phase can be destructive in the best sense; it's where the design gets picked apart, modified, and rebuilt based on real feedback. You have to be ready to adapt quickly, even if it means reworking significant parts of the project. The goal is to build something that genuinely meets users' needs—not just what you think those needs might be.

One of the key takeaways is to treat the prototyping phase as a learning experience. It's not only about validating your ideas but about discovering the unknowns that can push your design further. For instance, while testing with people who had visual impairments, we stumbled on the need for a more nuanced approach to screen reader compatibility. It wasn't just about making text audible—it was about creating an experience where auditory feedback matched the pace of user interaction without overwhelming them. This insight only came to light because we set up the testing strategy to engage deeply with diverse user experiences.

Ultimately, defining a strong testing strategy means designing your prototypes with intent and embracing the lessons that come from failures and unexpected outcomes. The more tailored your prototypes are to different user scenarios, the better you can refine the final product into something truly inclusive and impactful.

Making Testing Count: A Deep Dive into User Feedback

After defining the testing strategy, the next step is to dive into a comprehensive testing phase. Although it's impossible to test every potential scenario, the goal is to expose the solution to a wide range of users whose feedback can help predict how it will perform in real-world conditions. Accepting that there will always be limits due to time and resource constraints, the objective is to gain as much insight as possible by testing with a diverse pool of users who represent different backgrounds and needs.

One key approach for making testing more inclusive is to ensure diversity in user groups. In my lab, we often cross-share user bases between different projects, meaning one project's participants might also test another project's prototype. This forces each team to incorporate perspectives from varied and often excluded populations. Surprisingly, rather than overwhelming our beta users, this approach kept them engaged; they enjoyed being part of something new and having the chance to shape products that address broader accessibility issues. Their understanding of exclusion challenges made their feedback especially valuable and constructive.

For anyone looking to make testing as inclusive as possible, collaboration within the broader Equitable Design space is essential. Although this discipline isn't yet universally mainstream, it is far from being a solitary effort. Equitable Design is a growing movement driven by a collective of designers, researchers, organizations, and institutions dedicated to creating more inclusive, accessible, and impactful solutions. It is inhabited by a dynamic network of contributors—from individual designers and design agencies to local organizations, nonprofits, and public institutions—each working to address systemic inequities in their unique contexts.

The field is well-documented by pioneers who have shared their work to inspire and guide others. Books like *Design for Good* by John Cary (2017, Island Press) and *Design Thinking for the Greater Good* by Jeanne Liedtka and her colleagues

(2017, Columbia Business School Publishing) provide rich case studies of Inclusive Design in action. Annie Jean-Baptiste has become a leading voice on the topic through her role at Google and her book *Building for Everyone* (2020, Wiley), which outlines best practices for creating inclusive products. Beyond these prominent figures, countless local organizations and nonprofits are doing vital work every day to support marginalized communities. For example, I've had the privilege of collaborating with groups like Success Centers and the SF Tech Council in the Bay Area, who are actively bridging gaps in opportunity through Equitable Design initiatives.

Public institutions and government agencies are also pivotal players in the Equitable Design space. Their ability to impact large-scale change makes them invaluable partners in this work. For instance, the City of San Francisco's Office of Housing and Community Development has been an exemplary ally in advancing equitable solutions in housing and urban development. These organizations bring not only resources and networks but also deep knowledge of the communities they serve, enabling more impactful and sustainable design outcomes.

This collective effort demonstrates that Equitable Design is not merely a collection of isolated projects or initiatives. It is a movement rooted in collaboration and shared purpose, focusing on building a more inclusive world. By engaging with the wealth of knowledge and resources within this space, we can continue to push the boundaries of what Equitable Design can achieve, learning from each other's successes and challenges along the way.

Sharing access to diverse user groups can significantly enhance the depth and breadth of feedback received.

It's important to combine both qualitative and quantitative testing methods to get a full picture of how the solution performs. Qualitative methods might include user interviews, direct observation, and contextual assessments where the team demos the product or prototype in real-world settings, asking questions and capturing user reactions. This hands-on approach allows you to dig into why a feature might be confusing or frustrating.

On the quantitative side, we often set up self-explanatory prototypes, landing pages, or mockups that users can interact with independently. These prototypes are designed to gather data without requiring extensive instructions. For example, tracking how users navigate an interface can reveal whether certain features are intuitive, if buttons are overlooked, or if there are unexpected workarounds. This data helps identify patterns, such as a commonly missed button or recurring negative emotion during the user journey.

Documenting the testing phase is crucial. Although an individual piece of feedback may seem insignificant, when multiple data points start to align, they can reveal a pattern that requires attention. For example, in a past project, we kept noticing users hesitating at a specific step in the process. At first, it seemed like an isolated issue, but over time, as similar comments accumulated, we realized we needed to simplify that step to improve the overall experience.

Ideally, a well-executed testing phase won't take you back to the drawing board if co-design principles and thorough user observation were followed earlier. Instead, it should be about refining and optimizing the solution. This stage is where concepts start transforming into tangible realities, critical tweaks are made, and real excitement of solving a genuine problem starts to take shape.

Ultimately, a good testing phase requires the empathy of a designer who can listen to users and embrace uncertainty, combined with the analytical rigor of a data scientist. The effort put into this phase will pay off, as the insights gained will lead to a solution that resonates with users and achieves the intended impact.

Adapting to Company Timelines While Staying Agile

It's essential to acknowledge that equitable and inclusive solutions can't always be implemented overnight. Every company has its own workflows, schedules, and processes, and DEI initiatives need to fit within these established frameworks to ensure proper integration and scalability. If inclusivity efforts are treated as add-ons rather than embedded within existing processes, they can struggle to achieve long-term success.

Take, for instance, the project with the notebook company. We knew going in that we were looking at a one-year lead time due to the seasonal nature of the products and the demands of the supply chain. Orders for the back-to-school season needed to be placed months in advance, meaning we had to operate under these time constraints.

Of course, there was some frustration, knowing that it would take a full year before the impact of our efforts would be seen on store shelves. In a rapidly evolving space like DEI, this timeline comes with risks. Competitors could release a similar product, public perception could shift due to new headlines, or a new technology could emerge that changes the landscape entirely.

But there was also a silver lining: the extended timeline gave us more opportunities to refine the product and conduct further testing. We were able to connect with a broader group of users and collect more feedback, allowing us to make thoughtful iterations. This kind of continuous feedback loop ensured that when the product did launch, it would be as effective and impactful as possible.

Moreover, because we had explored alternative distribution channels outside of traditional retail, such as partnerships with occupational therapists and special education programs, we were able to test the product in smaller pilot programs. These pilots allowed us to gather feedback on a manageable scale, making it easier to incorporate changes without the pressure of a large-scale rollout. This smaller-scale testing gave us valuable insights and helped build momentum ahead of the product's broader launch.

Running these pilot programs is often a key step for any successful implementation. Starting at a smaller scale helps manage risks and lets you see how a solution is received in real-world conditions. It allows for fine-tuning based on real user experiences before committing to a full-scale launch. It's a critical step that not only informs the final iteration of the product but also builds confidence among stakeholders that the investment in Inclusive Design is paying off.

Although it may feel challenging to wait for the full impact to unfold, the time invested in preparation, testing, and iteration ultimately ensures that the solution is not just a one-off effort but a sustainable part of the company's offerings.

Aligning Teams for Inclusive Design Implementation

Reshaping and adjusting teams is a crucial step in implementing Inclusive Design. Although the design team naturally comes to mind first, the broader organization needs to be aligned with the principles of Equitable Design to truly make an impact. The design team should reflect diversity not just in background but also in skills and understanding of DEI principles. This diversity within the team supports the intention behind inclusive projects, bringing a range of perspectives that enrich the design process and lead to more innovative solutions.

I once invited a guest speaker to my class: Mahd, a remarkable gentleman leading product marketing for Google. Among his roles and responsibilities was a focus on improving the sense of belonging for both employees and users, with a special emphasis on faith inclusion. His work illustrated how deeply Google had committed to fostering a culture of belonging, supporting employees' varied identities, and being consistent with its DEI promises. It showed that Inclusive Design isn't just about products—it's about creating environments where everyone feels represented.

However, it's not enough to limit these efforts to the design team. The rest of the organization—operations, sales, marketing, and customer service—also needs to be aligned with the goals of Equitable Design. Once a product leaves the hands of the designers, it's the sales teams, customer service representatives, and marketing staff who directly engage with customers. If they aren't familiar with the concepts and the goals behind the new design, there can be a disconnect that impacts the product's success in the long run.

A strong example comes from a consulting project we did on neurodivergence with an alcoholic beverage company. The project was framed as a unique opportunity for innovation within the organization—a chance to challenge existing products while also raising awareness and educating key stakeholders about neurodiversity. As the project was introduced internally, several employees who identified as neurodivergent reached out on their own initiative, eager to contribute. Their involvement was driven by a combination of wanting their voices heard and acting as guardians to ensure that the process stayed authentic and produced meaningful results.

Their insights were invaluable, offering ideas they had accumulated over the years, highlighting personal pain points, and sharing remarkable initiatives they had observed elsewhere. For example, they pointed us toward adaptations made by a local museum to accommodate neurodiverse audiences at specific times of the week. Their involvement didn't just enhance the project—it became a way to pressure-test our ideas and ensure that we approached the design process thoughtfully. Knowing that no one on our team was neurodivergent, we relied on them to challenge assumptions, refine concepts, and guide the direction of our work.

They played a key role in pointing out false assumptions, rejecting ideas that didn't feel novel or impactful, and steering us toward approaches that resonated with their lived experiences. Their contributions sparked broader conversations within the company, raising awareness about DEI and fostering engagement among colleagues. Their presence also kept us accountable, pushing us to deliver our best work and approach the project with the rigor and sensitivity it demanded.

Their involvement ensured that the final outcome wasn't just informed by their experiences but was also shaped by their unique perspectives, making the project both meaningful and impactful.

Involving a broader group from within the organization also serves to build buy-in. When different teams are part of the process, they not only understand the new product better but also advocate for its success. This kind of company-wide engagement encourages a culture that pays attention to DEI topics, leading to better employee participation and morale and attracting talent who value inclusiveness in the workplace.

Ultimately, implementing inclusive solutions isn't just about designing the product right—it's about ensuring the organization as a whole is equipped to support and promote those solutions throughout its lifecycle. This alignment contributes to the sustained success of DEI initiatives and can foster a culture that continuously seeks to improve.

Notes

[1] StrategyCorps, n.d. Economic realities of retail checking in today's marketplace. [Online]. Available at: `https://strategycorps.com/discovery-center/articles/economic-realities-retail-checking-todays-marketplace` [Accessed: 15 January 2025].

[2] Sustainable Brands, 2024. Inclusive advertising boosts sales & brand value. [Online]. Available at: `https://sustainablebrands.com/read/inclusive-advertising-boosts-sales-brand-value` [Accessed: 15 January 2025].

[3] Kantar, 2024. Three quarters of consumers say inclusion and diversity influence their purchase decisions. [Online]. Available at: `https://`

www.kantar.com/company-news/three-quarters-of-consumers-say-inclusion-and-diversity-influence-their-purchase-decisions [Accessed: 15 January 2025].

[4] Statista, 2024. Customer attitude towards diversity and inclusion in advertising in the U.S. and UK in 2022. [Online]. Available at: https://www.statista.com/statistics/1331225/customer-attitude-diversity-inclusion-in-advertising-us-uk [Accessed: 15 January 2025].

[5] Florentine, S., 2020. IBM, HP partnerships with HBCUs boost diversity in tech. ChannelE2E. [Online]. Available at: https://www.channele2e.com/news/ibm-hp-partnerships-hbcus-diversity [Accessed: 15 January 2025].

[6] Black Engineer, n.d. IBM expands partnerships with HBCUs to build a diverse STEM workforce. [Online]. Available at: https://www.blackengineer.com/imported_wordpress/ibm-expands-partnerships-hbcus-build-diverse-stem-workforce-mp-1 [Accessed: 15 January 2025].

[7] Greenemeier, L., 2020. 5 Things to know about IBM's technology investment in historically black colleges and universities. IBM Research. [Online]. Available at: https://newsroom.ibm.com/archive-IBM-research?item=32437 [Accessed: 15 January 2025].

[8] Panasonic, n.d. Universal design products. [Online]. Available at: https://holdings.panasonic/global/corporate/universal-design/products.html?utm_source=chatgpt.com [Accessed: 15 January 2025].

[9] Ng, K., 2023. Japan population: One in 10 people now aged 80 or older. BBC [Online]. Available at: https://www.bbc.com/news/world-asia-66850943 [Accessed: 15 January 2025].

[10] Maximize Market Research, 2024. Assistive devices market: global industry analysis and forecast (2024–2030). [Online]. Available at: https://www.maximizemarketresearch.com/market-report/assistive-devices-market/152687 [Accessed: 15 January 2025].

[11] BBC, 2015. Google apologises for Photos app's racist blunder [Online]. Available at: https://www.bbc.com/news/technology-33347866 [Accessed: 15 January 2025].

[12] Simonite, T., 2018. When it comes to gorillas, Google Photos remains blind. *Wired* [Online]. Available at: https://www.wired.com/story/when-it-comes-to-gorillas-google-photos-remains-blind [Accessed: 15 January 2025].

[13] Kayser-Bril, N., 2020. Google apologizes after its Vision AI produced racist results. AlgorithmWatch [Online]. Available at: https://algorithmwatch.org/en/google-vision-racism [Accessed: 15 January 2025].

Scaling Up: From Local Initiatives to Global Impact

My work has largely been centered on the inception phase of Inclusive and Equitable Design, primarily because the field is relatively new. The most critical work needed to be done at the very beginning, when it was most necessary.

However, I have always understood that the ultimate goal of Equitable Design is to scale its impact and create a ripple effect. From the start, I focused much of my efforts on engaging with large, global companies. Although I deeply value what startups contribute to society—especially having lived in Silicon Valley and being an entrepreneur myself—I believe that large organizations, which already impact millions of people, can have an even greater influence. These companies employ large, diverse pools of talent across communities and reach an enormous audience with their products.

If we can make these organizations more inclusive and foster positive conversations about equality, even small changes can lead to massive results. Improving inclusivity by just a few percent within the Fortune 500 companies would have a much larger reach than any single university, religious institution, or political entity could ever achieve.

This chapter explores how we can scale the work that Inclusive Designers initiate and propagate their findings, commitments, and solutions to achieve widespread impact.

Orchestrating Change

After the events surrounding George Floyd, many American companies turned inward, investing in internal transformations and committing millions to hire a more diverse workforce. Community support groups began to emerge in all directions, and executives delivered inspiring speeches on the importance of equality in the workplace.

This was certainly a strong starting point. It's often easier to have these conversations internally, testing and learning within the company before making external commitments.

Following are a few recommendations I've been advocating to my clients over the past few years. These go beyond just aggressive policies aimed at increasing workforce diversity and fostering inclusive work environments—both of which remain top priorities for business leaders.

Leading with Vulnerability: The Key to Scaling Inclusive Design

One of the biggest challenges in promoting Inclusive Design lies in leadership roles. Leaders hold significant influence over both their organizations and stakeholders, but they are also under intense scrutiny. The reality is that most leadership positions are still occupied by white, older, highly educated males, making it difficult for them to authentically lead conversations around diversity and inclusion. This disparity, rooted in historical privilege, often places leaders in a delicate position as they try to "walk the talk" on issues they may not fully relate to or understand.

As I've learned from personal experience, this is where it becomes crucial to understand that tackling inequality is a responsibility for everyone, especially those in privileged positions. They must work to actively address and help rectify disparities.

To effectively scale Equitable Design, one of the key transformations needed is to turn leaders into ambassadors and role models for inclusion. Training them and offering opportunities for growth in their leadership skills enables them to drive changes that will cascade across their organizations. Inspirational speeches and public pledges can initiate change, but without authentic, consistent actions, they often fall short.

What may seem counterintuitive is that a leader's ignorance can actually be their best ally in this journey. I've always advocated for vulnerability as a defining principle of inclusive leadership. By showing up with the humility to admit that they don't know everything and are willing to listen and learn from others, leaders can model a positive approach that invites others to follow suit.

I've seen firsthand how detrimental it can be when leaders, out of fear of appearing weak, assert opinions that misinterpret the needs of their workforce.

It's far more damaging to deliver an exclusive opinion than to admit, "I don't know." This humility fosters a space where leaders can be called out without repercussions, and employees feel comfortable voicing their concerns and ideas. This openness builds trust and contributes to creating a genuinely inclusive environment.

I've personally embraced vulnerability as a CEO. At one point, a close collaborator—a woman of Asian descent deeply committed to social justice— expressed her frustrations. She felt that her ideas were consistently dismissed, whereas those from male colleagues gained more traction. My initial reaction was defensive, as from my perspective, her critique seemed unfair. But because she approached the issue with honesty and respect, and because I was committed to practicing vulnerable leadership, I was able to reflect on her words.

She pointed to an example where she had early on suggested implementing a badging system for clients and students. At the time, I wasn't fully sold on the idea and let it drop. Months later, after discussing the concept again with a male colleague, we refined it and decided to move forward with implementation. From her viewpoint, her idea wasn't heard until it was presented by a male colleague. From my side, I saw it as a collective evolution. But what mattered was acknowledging her perspective, crediting her contributions, and recognizing how unconscious bias played a role.

This experience taught me a lot about how small gestures of vulnerability and openness can lead to greater team empowerment. The conversation made her feel heard and valued, and I became more conscious of my biases. The rest of the team also saw the positive outcome, encouraging them to speak up and embrace vulnerability in their leadership roles. And all of this took just a few conversations and the right mindset from both sides.

But unfortunately, this is far from the most common scenario. In many cases, the conversation would likely have ended with both sides firmly standing their ground, arguing their perspectives, and further reinforcing tensions and misunderstandings. This latter situation is probably all too familiar to most readers.

Vulnerability as a leadership quality is no trivial matter. Society, education, and the media have long pushed leaders to exude strength, competence, and assertiveness. Many publicized leaders still follow this model, but I believe we need to unlearn and relearn this mindset to foster environments where Inclusive and Equitable Design can truly thrive.

Meet Them Where They Are

Leaders, by definition, are busy individuals juggling multiple high-stakes responsibilities. From my experience, the most effective approach is to meet them where they are—in terms of both mindset and context. Online diversity, equity, and inclusion (DEI) training, although well-intentioned, tends to have limited impact at this level. What truly resonates is peer connection and exposure to real-world examples.

A great illustration of this came with the rise of CEO Action for Diversity & Inclusion, a platform that rapidly gained traction to become the largest CEO-driven initiative committed to advancing workplace diversity. The CEO Action for Diversity & Inclusion initiative invites CEOs from companies of all sizes to sign a pledge committing to advance diversity and inclusion in their workplaces. Launched in 2017, the pledge outlines specific actions that signatories will undertake, such as fostering a trusting environment where all ideas are welcomed and encouraging open dialogue about diversity and inclusion issues. The platform serves as a collaborative space for CEOs to share best practices, resources, and progress on their diversity initiatives. Signatories are encouraged to implement unconscious bias training, engage in difficult conversations, and create a culture of accountability within their organizations.

In July 2020, the platform reported 1,100 signatories, 1,000 concrete actions shared, and 89% of participating organizations claiming to have implemented unconscious bias education initiatives.[1] Fast-forward a few years, and the platform boasts over 2,500 signatories, including CEOs from major companies like Accenture, Adobe, Deloitte, Procter & Gamble, and Cisco.[2, 3]

The initiative's success lies in its framing of inclusion and diversity not merely as ethical imperatives but as business-critical goals. CEOs weren't just signing up—they were drawn in by the peer-driven, high-visibility nature of the movement. The collective commitment of their peers sent a powerful message: inclusion was no longer a "nice-to-have" but a business necessity essential to staying competitive in an evolving market. Each company could share some actions taken, and a mandatory point to be shared was the measurable outcome of the action.

However, it's important to critically assess the scope and depth of these reported actions. Most initiatives focus on providing unconscious bias training, advocating for transparency, improving team dynamics, and introducing diversity criteria in hiring. Although these steps are valuable, the actions reported are self-declared by the companies themselves, making it difficult to measure systemic impact.

CEO Action has undeniably served as a powerful platform for sparking the adoption of DEI practices and fostering dialogue, as well as a strong communication tool for companies. Yet the broader, long-term systemic repercussions of these efforts remain to be seen. It's a promising start but sustained accountability and measurable outcomes will determine whether the initiative achieves its transformative potential.

Beyond peer influence, it's essential to appeal to what leaders prioritize: economic growth, reputation, and shareholder trust. For this reason, the business case for Inclusive Design is crucial. Highlighting how social responsibility is now a key criterion for investment funds or demonstrating how Inclusive Design leads to more innovative products can help initiate conversations and bring leaders on board.

Sometimes a Little Nudge Goes a Long Way

Even after convincing leaders to embrace inclusive leadership, it's not something that develops overnight. Vulnerable and inclusive leadership takes practice, and often, a little nudge can go a long way in embedding these values into their daily routines.

One example comes from a project we worked on with a large software company. We integrated inclusive messaging into the company's project management tool, encouraging team leads to acknowledge and credit all contributors when a project was completed and to highlight how diversity and teamwork led to successful outcomes. These small but meaningful nudges acted as reminders to reinforce inclusive practices in real time, helping leaders develop habits that eventually become second nature.

Leaders who embrace this vulnerable approach, allowing for reminders and small course corrections, ultimately foster more engaged, productive, and resilient teams. In the long run, this positive environment pays off—driving both individual and organizational success.

Aligning Incentives: Building Accountability and Engagement for Inclusive Growth

Another aspect that often gets overlooked in large organizations is the chain of incentives—how to ensure that everyone is aligned and empowered to contribute meaningfully to key objectives.

I like to explain this by drawing parallels to the U.S. healthcare industry. In this system, insurance companies are incentivized to limit coverage and minimize reimbursements to protect their financials. Meanwhile, doctors and healthcare providers are driven to prescribe more treatments and services, increasing their revenue and allowing them to invest in their practice. At the same time, patients are incentivized to minimize their use of healthcare services to avoid financial burdens. This misalignment creates a situation where each party is working against the interests of the others.

In contrast, HMO programs take a different approach by combining insurance and healthcare delivery under one entity. Here, the incentive is to keep patients healthy to avoid costly treatments, benefiting both the patient and the insurer. The focus shifts to preventive care and regular check-ins, aligning the interests of the patient, provider, and insurer in a way that promotes better outcomes for everyone.

This model is reminiscent of the narrative surrounding traditional Chinese medicine, where patients pay their doctor as long as they are healthy but stop payments when they fall ill—effectively incentivizing doctors to focus on prevention and overall well-being. Although the concrete evidence supporting the existence of such a compensation model in traditional Chinese medical practice is limited, the idea of compensating professionals based on their clients' outcomes is inspiring and aligns well with the preventive ethos of traditional Chinese medicine.[4]

We can apply these ideas to large organizations by examining how incentives are structured within the company. By aligning incentives across departments and ensuring that everyone is working toward the same goals, we can create a more unified, effective organization.

A transformative example from my experience as a sustainable designer involved working with one of the largest cosmetic brands in the world. We were tackling a problem we called the "zero-use product"—items that are designed, manufactured, shipped, but never sold or used. This issue is particularly prevalent in industries like fashion or food, where products have limited shelf lives due to sanitary or fashion trends.

In our case, we were dealing with cosmetic products that were distributed globally and stocked in stores but eventually destroyed because they were never purchased. This was problematic for several reasons: environmentally, these products couldn't be recycled or reused, wasting the resources spent on production and shipment. From a business standpoint, the company not only lost money on the unsold products but also had to pay to dispose of them.

Through our research, we discovered the root of the issue: supply managers were incentivized to avoid product shortages at all costs. Their bonuses were tied to ensuring that no products were out of stock when there was demand. This system encouraged overstocking to avoid any missed sales, which in turn exacerbated the waste problem.

By revising the incentive structure—encouraging managers to aim for leaner stock levels—we were able to significantly reduce the issue. This not only solved a major environmental and financial problem but also unlocked the full potential of the managers' expertise and creativity, as they were now incentivized to find more efficient solutions to seemingly contradictory goals. This reflection offered a more systemic perspective on the way the supply chain was structured in the company and offered managers a wider view of their responsibilities and their ability to sustain and grow the business. Their actions and importance were no longer limited to the level of stock but related to the overall precision of the supply chain and the ability to have the right number of items at the right place and the right time.

The recommendations I consistently provide start with defining the end goal: understanding what the overarching objective is and how each person's role contributes to achieving it. Once this is clear, the next step is to cascade these objectives down to individual levels, ensuring that everyone in the organization has a clear sense of their role in the bigger picture.

When it comes to incentives, it's important to focus not only on financial rewards but also on more nuanced incentives that motivate different types of contributors. Financial bonuses are often the most obvious way to incentivize people, rewarding those who meet or exceed their targets.

For instance, designers and product managers can be incentivized through bonuses based on their performance against accessibility metrics—a critical measure in ensuring Inclusive Design practices. Managers, on the other hand, can receive bonuses tied to qualitative and quantitative feedback related to DEI within their teams. This approach not only rewards results but also encourages ongoing improvements in workplace culture. For sales and customer-facing roles, incentives can be tied to the diversity of their audience and the feedback they receive, ensuring that they are engaging with a broad, representative customer base.

In a later section, we'll delve into KPIs (Key Performance Indicators) and metrics to explore how to articulate these goals in ways that are both ambitious and meaningful. This ensures that the incentive structure is clear, motivating, and aligned with the company's larger mission.

One such lever is the opposite of bonuses: consequences. Certain things must be non-negotiable and, when breached, should come with clear consequences. For example, if a salesperson fails to meet their targets or a consultant neglects to show up for a client engagement, there need to be repercussions, ranging from light sanctions to, in extreme cases, job termination. The goal here isn't to create an atmosphere of fear but rather to ensure that the stakes are appropriately aligned with the organization's expectations.

Similarly, discriminatory behavior—whether in interpersonal interactions or in the design of products or services—must also carry consequences. Discrimination based on race, gender, or sexual orientation is punishable by law, and this should be enforced rigorously, not just in relationships between colleagues and clients but also in the way products and services are conceived. A product or service that purposely excludes certain groups should be treated as an act of discrimination. DEI needs to be taken as seriously as any other business imperative, and the consequences for failing to meet these standards should reflect its importance.

In her insightful TED Talk titled "How to Get Serious About Diversity and Inclusion in the Workplace",[5] Janet Stovall outlines a practical, results-driven framework for driving real change in DEI. She emphasizes the importance of defining real problems, real numbers, and real consequences as the foundation for making meaningful progress.

Stovall shared a personal example from her time as a student when she initiated Project '87 at Davidson College. The project set specific, measurable goals for the institution: within three years, the school needed to enroll 100 Black students, hire 10 Black professors, create five Black Studies classes, and appoint one Black dean. Although these targets may not seem revolutionary, the key difference was the accountability embedded in the challenge. Stovall and her peers demanded that if the college failed to meet these goals, its commitment to diversity would be openly questioned. This clear combination of measurable targets and accountability ensured that the goals weren't just aspirational but achievable and enforceable.

Stovall suggests that businesses can adopt a similar approach by clearly identifying the specific diversity and inclusion challenges they face, whether related to workforce representation, product accessibility, or addressing underserved demographics. Once the issues are identified, companies should establish data points—quantifiable measures to track progress—and define the consequences if these goals are not met. This ensures that DEI is treated as a central and critical priority within the organization.

By integrating this structured approach, companies can move beyond superficial diversity efforts and create systems that foster genuine and sustained inclusion.

On a more positive note, some simple but effective incentives can make a significant impact in fostering engagement and commitment toward DEI goals. Recognizing and giving credit to employees for their dedication and contributions, rewarding both workers and clients for their creativity, and offering opportunities for individuals to champion initiatives can go a long way. These small acts of recognition not only motivate people but also signal that their efforts in advancing inclusion are valued.

Moreover, DEI-related roles often carry significant reputation and recognition. These positions are closely linked with values of progress, social advancement, and justice, making them highly attractive for individuals who want to align their work with meaningful impact. Such roles and initiatives are often seen as valuable additions to a resume, as they signal an individual's commitment to creating a more inclusive environment—qualities that are increasingly sought after by organizations in today's job market.

These efforts can create a positive feedback loop, where employees feel valued for their contributions, which in turn promotes greater engagement and further commitment to DEI efforts across the organization.

This discussion raises the contentious topic of affirmative action, referred to in French as *discrimination positive*, or "positive discrimination." Framed this way, it might seem like an oxymoron. Affirmative action encompasses policies and practices aimed at addressing historical inequities and discrimination by creating opportunities for underrepresented groups, particularly in areas such as education, employment, and government contracts. Its goal is to level the playing field by taking into account factors such as race, gender, or socioeconomic status to foster diversity and representation.

For instance, in the United States, affirmative action policies in higher education admissions have been utilized by institutions like Harvard University to boost enrollment of students from historically marginalized racial and ethnic groups. Similarly, in the workplace, companies like IBM have implemented targeted recruitment and training programs to enhance diversity and create equitable opportunities for women and minority candidates in leadership roles.

These examples illustrate how affirmative action can operate across various domains to promote inclusion and address systemic disparities.[6, 7]

Although these efforts are well intentioned and aim to uplift individuals affected by systemic inequities, they also present two significant challenges. The first is that affirmative action can disadvantage those who do not belong to the supported groups, potentially creating another form of discrimination. For example, as seen in the context of Janet Stovall's discussion, individuals from other ethnic backgrounds may feel excluded from opportunities like university admissions due to these policies. This can foster frustration, a sense of injustice, and, over time, resentment toward those perceived as benefiting unfairly—risking an undesirable cycle of division. Some of the institutions that applied affirmative action had to face trials and reputational challenges because of those practices.

The second challenge lies in evaluating individuals based on the groups they belong to rather than their unique qualities. This type of selection can have long-term consequences, such as fostering imposter syndrome among beneficiaries and resentment among those who feel their true value is overlooked. Instead of being recognized for their abilities, individuals may feel reduced to a token of diversity, which can undermine both confidence and trust.

These limitations do not imply that initiatives like affirmative action are counterproductive. Rather, they underscore the inherent difficulty in designing perfect interventions. Efforts to change systemic inequities often involve trade-offs and unintended consequences. Recognizing this is vital for fostering an approach to Inclusive and Equitable Design, as well as DEI efforts. There is no single solution, and most interventions come with their own set of challenges and side effects. Transforming entrenched systems requires bold actions and sacrifices—not just for the direct results but for the conversations and shifts in perspective they inspire.

The recommendation I often make to my partners and myself is to remain focused on the ultimate goal and ensure that our actions align with that vision. In Janet Stovall's context, her ultimate aim was to uplift the Black community through greater access to education and a more authentic educational experience. Education, like knowledge, retains its value as it is shared. Although this goal is inclusive and equitable, it is important to acknowledge that the path to achieving it may involve trade-offs and sacrifices.

Supplier Diversity

Supplier diversity has become an increasingly important strategy for driving inclusivity on a larger scale, gaining significant traction in recent years. Beyond making changes internally, companies can make a positive impact through their external partnerships, including suppliers and collaborators.

Large corporations spend millions annually on goods and services to keep their operations running, which gives them the power to drive ethical change by intentionally sourcing from businesses that are often underrepresented. By prioritizing supplier diversity, these companies contribute to more equitable economic opportunities for businesses owned by women, people of color, veterans, people with disabilities, and other marginalized groups.

For example, companies like Coca-Cola and Walmart have long-standing supplier diversity programs, setting measurable goals to increase the diversity of their supplier base. This not only helps uplift underrepresented businesses but also brings innovation and ensures that the company's supply chain reflects the diverse customer base it serves.

To make supplier diversity effective, procurement departments should set clear metrics to evaluate suppliers and communicate transparent policies about diversity. Tracking the percentage of spending with diverse suppliers and offering mentorship to smaller businesses are effective ways to push for real change. This ripple effect supports economic growth in marginalized communities while promoting inclusive business practices.

Additionally, in my experience, companies that regularly survey their suppliers on DEI practices during the onboarding process help push these important conversations forward. This encourages suppliers to reflect on their own practices and improves their commitment to diversity, ultimately strengthening the entire supply chain's DEI initiatives.

Practical Metrics for Measuring Impact: Ensuring Tangible Outcomes in Inclusive and Equitable Design

A critical aspect of Equitable and Inclusive Design is the work around defining metrics. These metrics form the backbone of a project, and this process can easily belong to any chapter of the design journey, as it's something that needs to be addressed and evaluated at every step along the way.

Equitable and Inclusive Design are still relatively new concepts, especially in how they are theorized and applied. They also deal with elements that can be more challenging to quantify or get immediate feedback on. For instance, how do you measure something as intangible as a sense of belonging? How can you gauge the level of equality between different groups in a reasonable time frame? These are more complex than traditional business or technical metrics like social media engagement and a company's financials.

Nevertheless, these success metrics are essential because they provide direction and ensure everyone involved keeps the ultimate goals in mind. They serve as a reminder of what we are aiming to achieve and help maintain motivation throughout the project as they point toward the end game.

I always encourage my teams to think about KPIs early in the process. This is especially helpful in projects where the project owners might have a strong intention but lack a fully formed sense of the outcome. Defining KPIs brings clarity.

For instance, when working on a project to create more inclusive celebrations for neurodiverse individuals with a beverage company, what are we actually striving for? Are we aiming for higher attendance at events? Do we want attendees to have a better experience? Should they feel more involved in the celebration? How do we measure that? Should we track audience size, survey participants to assess neurodivergent representation, or measure how long people stay at the event? Perhaps we could track how many positive interactions attendees have or even how many drinks they consume. All these questions are crucial in shaping the direction of the project, defining what resources are needed, and structuring the entire approach.

I've created a two-step framework to help guide this process:

1. Identify easy-to-measure metrics.

2. Consider long-term goals.

Identifying Easy-to-Measure Metrics

The first step is to identify easy-to-measure metrics that can help shape the project and inform the design of the product or service. This step builds on the Benefit Matrix introduced earlier in the book (see Figure 8.1).

The matrix at the center helps companies and leaders determine where to focus their efforts and identify where the opportunities lie. It has two axes: the horizontal axis ranges from reducing disadvantage to empowering communities. On one end, the goal is to improve accessibility to a product or service by enhancing ergonomics or user experience breaking down barriers for target users. On the opposite end, empowering communities implies that using the product or service can help underserved groups gain ground and improve their position over time.

The vertical axis runs from universal design to personalized design. Universal design focuses on creating a product or service that anyone can use, which may seem like the most inclusive approach. However, it can sometimes lead to solutions that don't fully meet the specific needs of certain groups. Personalized design, on the other hand, tailors solutions to specific communities, offering more targeted benefits but potentially sacrificing broader appeal. Deciding on the right balance for each project is a crucial trade-off to consider.

Once you've identified which quadrant of the matrix the project falls into, you can begin selecting metrics to ensure the goals are being met. I've developed suggested metrics models for each of the four directions in the matrix. Although these are tailored to specific contexts, they can serve as a useful source of inspiration for a wide range of projects.

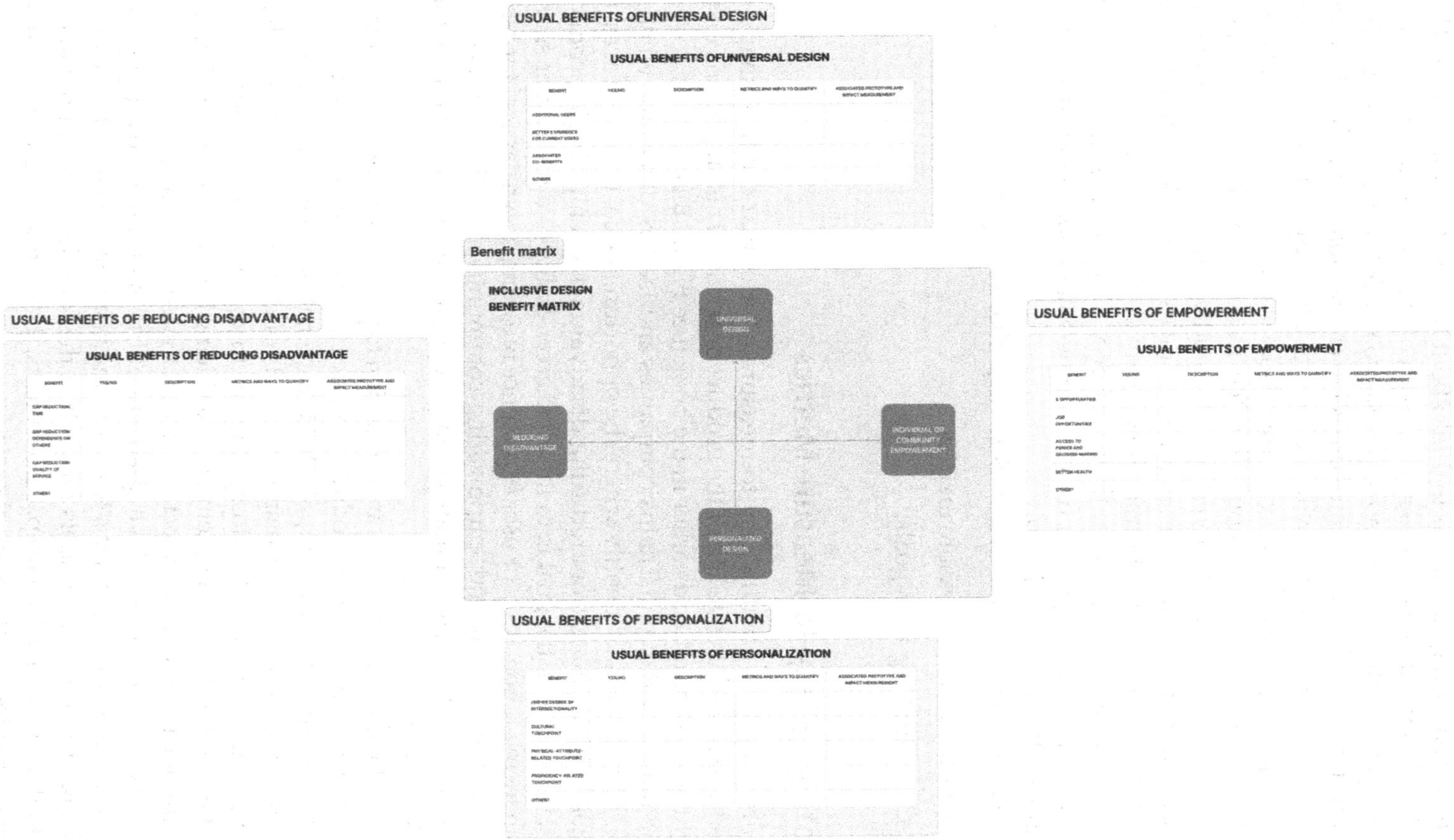

Figure 8.1: Metrics guidelines for the Benefit Matrix

Let's start with the Reducing Disadvantage direction. As you can see in the table in Figure 8.2, the goal here is to identify the benefit we are aiming for, essentially defining the specific problem we are solving. For each of these benefits, we need to assess whether the new design addresses the issue. This involves describing how the design will work, quantifying the impact, and determining what prototype or testing mechanisms can be used to measure this positive change.

USUAL BENEFITS OF REDUCING DISADVANTAGE

KEY BENEFIT OF INCLUSION	IS THIS ACHIEVED?	HOW THIS BENEFIT WORKS	WHAT METRICS TO MEASURE IMPACT?	EXAMPLES & PROTOTYPES
GAP REDUCTION: TIME				
GAP REDUCTION: DEPENDENCE ON OTHERS				
GAP REDUCTION: QUALITY OF SERVICE				
OTHER BENEFITS				

Figure 8.2: Metrics guidelines for reducing disadvantage

For the Reducing Disadvantage half of the matrix, here are some common benefits we've identified based on past projects:

▪ **Reduction in time spent**: A key benefit can be reducing the time it takes for underserved users to complete certain actions. For instance, in an authentication project for blind users, the new design significantly cut down the time needed to log in—saving 27 seconds, measured through testing that compared the current system with the new geolocalized version.

▪ **Reduction of dependence on others**: Another metric is the decrease in reliance on help from family, friends, or caregivers. This was a major focus in an inclusive banking project where we redesigned the user journey to make financial tools more accessible, minimizing the need for financial literacy support. The success was measured by the reduction in times help was needed each month, dropping from 10 instances to just 1 with the new solution.

▪ **Quality of service**: Will the target audience still receive a lower-quality experience compared to others? In a project aimed at improving the celebration experience for neurodivergent individuals, we created quieter, sensory-friendly pods in bars. This improved the experience by reducing the overwhelming noise and sensory stimuli. To measure success, we used qualitative surveys and monitored the average time spent in bars before and after the change.

These examples highlight how identifying the right benefit, quantifying the improvement, and using appropriate testing mechanisms can ensure we're reducing disadvantage effectively.

When focusing on empowering communities, I suggest four key metrics for success based on past experiences (see Figure 8.3).

USUAL BENEFITS OF EMPOWERMENT

KEY BENEFIT OF INCLUSION	IS THIS ACHIEVED?	HOW THIS BENEFIT WORKS	WHAT METRICS TO MEASURE IMPACT?	EXAMPLES & PROTOTYPES
INCOME-GENERATING OPPORTUNITIES				
JOB OPPORTUNITIES				
ACCESS TO POWER AND DECISION-MAKING				
BETTER HEALTH OR LIFESTYLE				

Figure 8.3: Metrics guidelines for empowering communities

- **Income-Generating Opportunities**: Are we creating paths for underserved communities to achieve financial growth? For example, in a social-media-enhancement project aimed at supporting Black-owned businesses, we sought to improve their visibility. This increased visibility directly contributed to business growth and financial gains. Measuring success here involves simple data points, such as tracking revenue growth through company accounting or using referral systems to trace the source of generated revenue.

- **Job creation in underserved communities**: Another critical metric is whether the solution creates job opportunities for communities in need. For example, in a project revamping the hiring process for a Champagne company, we made it significantly easier for individuals with disabilities to apply and be hired. By simplifying the application process and encouraging organizations to rethink their talent needs, we were able to open up job opportunities for underrepresented candidates. The impact here is easy to measure through HR data, tracking hires from the target community.

- **Access to power and decision-making**: Does the solution give the target audience greater access to power and decision-making roles within their community or organization? In our work with an insurance company seeking to expand its talent pool, some of the newly created positions carried high levels of responsibility. Filling these roles with candidates from the new talent pool directly increased their access to decision-making power. Tracking this can involve monitoring promotions, leadership roles, or the number of diverse voices in strategic positions.

▪ **Improving health**: Health is often a critical factor in community empowerment, as poor health can be a massive barrier to personal and economic development. For instance, chronic diseases limit people's ability to learn, work, and participate in society. In a project with a pharmaceutical company, our goal was to build trust in the healthcare system among minority communities, which we anticipated would lead to better health outcomes and greater empowerment. Success can be measured through surveys on trust in healthcare, as well as medical visit logs to track improvements in health access and outcomes.

These four metrics provide tangible ways to assess whether a project is truly empowering the communities it targets.

With universal design benefits, the most evident ways to measure are around the number of users and the quality of the experience (see Figure 8.4).

USUAL BENEFITS OF UNIVERSAL DESIGN

KEY BENEFIT OF INCLUSION	IS THIS ACHIEVED?	HOW THIS BENEFIT WORKS	WHAT METRICS TO MEASURE IMPACT?	EXAMPLES & PROTOTYPES
ADDITIONAL USERS				
BETTER EXPERIENCE FOR CURRENT USERS				
UNINTENDED POSITIVE IMPACTS				
OTHER BENEFITS				

Figure 8.4: Metrics guidelines for universal design

▪ **Additional users**: How many additional users can we bring to the product or service? Thinking of additional users involves calculating the net count—considering both new users gained and any existing users who might be lost. The aim of universal design is to serve as many people as possible, so the net increase should be substantial. We can approach this metric from two angles: either from a potential standpoint, where we estimate the size of the newly serviceable population, or from an actual traction perspective by examining the product's client base or sales data. For our authentication project, we made initial assumptions based on public data about the average percentage of people excluded from the current service due to impairments. We also reviewed how many users worked in various industries and adjusted the percentage based on their specific needs. For example, graphic design and aviation sectors have lower populations of visually impaired individuals, but these industries were significant clients of our partner. By applying these adjusted percentages to the overall user base, we arrived at a rough estimate of how many more people could be included in the new service. Although these estimates aren't exact, they provide a clear sense of the project's potential scope. This data not only informs the business case but also helps align scaling and deployment strategies. Ultimately, having a solid understanding of potential and real user numbers helps in determining both the reach and impact of a universal design solution, ensuring that it meets its inclusivity goals.

- **Better experience for current users**: The aim is to ensure that the design improvements benefit new users without compromising the experience for those already familiar with the product or service. In the case of the Champagne brand, we restructured the hiring process to better accommodate people with disabilities. However, this new process also proved advantageous for nondisabled users. The emphasis on working conditions and flexible working hours resonated with a broader range of candidates who needed a better balance between their personal and professional lives. For instance, a candidate who was a caretaker for a family member found the ability to customize their job search based on their schedule incredibly useful. This not only saved them time but also provided an enhanced job-seeking experience. In terms of measurement, this feedback can be both qualitative, like the example above, or quantitative, where data is available. For instance, in the authentication project, the improved process saved all users 27 seconds per login—a small yet significant time-saving that benefits everyone.

- **Unintended Positive Impacts**: The concept of co-benefits looks beyond the primary user group to determine if the design improvements are positively affecting additional stakeholders. For universal design to be truly universal, it should benefit three key groups: the target audience, the existing users, and other stakeholders who may be impacted. In the Champagne project, not only did the redesigned process make hiring more inclusive for people with disabilities, but it also had far-reaching effects for the employer and the wider workforce. By adopting more inclusive HR practices, the company sparked a shift in workplace culture, leading to broader changes in how they approach talent acquisition and employee engagement. These co-benefits often provide unexpected but valuable results, demonstrating how a universal design approach can improve the overall organization or community even beyond the initial target users.

By focusing on both direct and indirect benefits, universal design ensures that improvements ripple throughout the organization, enhancing experiences for a variety of stakeholders and making the solution as inclusive as possible.

As shown in Figure 8.5, the final area of the benefits matrix focuses on measurable benefits for personalized solutions, with the goal being precision in addressing the specific needs of the target audience:

USUAL BENEFITS OF PERSONALIZATION

KEY BENEFIT OF INCLUSION	IS THIS ACHIEVED?	HOW THIS BENEFIT WORKS	WHAT METRICS TO MEASURE IMPACT?	EXAMPLES & PROTOTYPES
APPLIED INTERSECTIONALITY IN DESIGN				
INTENTIONAL CULTURAL TOUCHPOINT				
INTENTIONAL PHYSICALLY ACCESSIBLE TOUCHPOINT				
INTENTIONAL LOW SKILL-BASED TOUCHPOINT				

Figure 8.5: Metrics guidelines for personalized solutions

- **Intersectionality**: This metric ensures that we are intentional in designing highly personalized solutions by accounting for the complexity of users' identities. Intersectionality, often discussed in theoretical frameworks around DEI, tends to be underutilized in practical applications. In our projects that focus on personalized design, we measure intersectionality by testing the solution against personas with distinct, layered identities. For example, in a social media project aimed at increasing visibility for businesses owned by people of color, we specifically targeted business owners of color who could benefit from higher exposure. We treated this as three degrees of intersectionality: race, profession, and market needs. The more specific we can be in defining these criteria, the more accurately we can tailor the solution. We measure intersectionality by defining the target audience and identifying the combination of characteristics that make them unique, thus ensuring that the solution resonates deeply with them.

- **Cultural touchpoints**: This focuses on how deliberate we've been in embedding cultural elements that may not directly influence functionality but help users feel that the solution was designed with them in mind. In a project with a fintech platform, for example, we redesigned the user interface to reflect the local culture for each key geographic region.

We measured cultural touchpoints by counting specific markers: local symbols, appropriate color schemes, fonts, language choices, and even the layout of the site's components. By cataloging these elements, we ensured that the design felt authentic and relevant to each region. This metric pushed us to make the product more culturally responsive and personalized.

- **Physical touchpoints**: These are especially important in projects designed to remove physical barriers for people with disabilities. When cultural inclusion isn't the main focus, physical accessibility becomes paramount, especially in hardware solutions or space design. For example, consider a project designed to improve the experience of people with visual impairments in a subway station. In this case, we multiplied physical touchpoints: tactile markers on the platform, braille signage, audio announcements, and ramps for easier navigation. In a similar vein, a project we conducted with a beverage company aimed to make celebrations more inclusive for neurodiverse people. We incorporated multiple physical touchpoints, such as quiet zones, fidget coasters, and tactile design elements throughout the space. Just as with cultural touchpoints, we counted these physical touchpoints to ensure they were consistent, effective, and accessible to the target audience.

- **Low skill-based touchpoint:** This metric evaluates how well a design leverages users' existing skills, knowledge, and learned behaviors to create a more intuitive experience. Instead of forcing users to adapt to entirely new systems, skill-based touchpoints ensure that designs feel familiar, accessible, and easy to use by building on what users already know. For example, in a project aimed at improving digital literacy among older adults, we incorporated touchpoints inspired by familiar offline experiences, such as using icons that resemble real-world objects (e.g., a mailbox for email, a folder for file storage). Similarly, in a professional training platform, we designed the interface to mirror common workplace software, reducing the learning curve for users transitioning from traditional office environments. By identifying and reinforcing skill-based touchpoints, we ensure that users feel confident and capable rather than overwhelmed, making the experience more seamless, engaging, and effective.

This approach to personalized design ensures that we are not only improving accessibility but also creating experiences that are deeply meaningful and relevant to the specific needs of the users, whether through intersectionality, cultural sensitivity, or physical accessibility.

All these metrics and indicators are meant to assist designers in identifying simple, measurable KPIs that can be applied throughout the design process. Although not exhaustive, they provide a framework for converting the inclusive and equitable intentions of a project into tangible, quantifiable outcomes. The goal here is to rationalize and quantify improvements, ensuring that the project stays focused on delivering real benefits to the target audience. These metrics also act as a check, preventing the team from getting so caught up in the excitement of the solution that they lose sight of the core purpose: serving the audience better.

However, it's important to note that these metrics are incomplete on their own. Relying solely on them won't guarantee the long-term success of the project. They serve as immediate indicators and are useful in the design phase, but a more holistic approach is needed for sustainable and impactful results. A long-term strategy, one that includes ongoing user feedback and iterative improvements, will be essential to ensure that the design continues to evolve and serve its intended community effectively over time.

Long-Term Metrics: Guiding Systemic Change in Inclusive and Equitable Design

The second step in the metrics process involves considering long-term goals. As discussed earlier, although some metrics can be measured relatively easily using the benefit matrix, others may be much harder to quantify clearly or within a reasonable time frame. Yet these long-term metrics are critical in establishing a clear future direction and setting an ambitious goal. Inclusive and Equitable Design is a long-term endeavor, and the effects of innovations in this area should be measured over time, not just in immediate results.

An example that has always resonated with me comes from my experience growing up in South Africa during the end of apartheid. The country faced an overwhelming challenge in trying to create an equitable society after decades of segregation. One thing that struck me was the widespread acknowledgment by both political leaders and citizens that change wouldn't happen overnight. They understood that it would take several generations before the impact of equitable policies could be fully seen. The inertia inherent in transforming a deeply unequal system was well understood. For instance, it was accepted that it would take a generation of Black children to attend the best schools, gain experience, and rise to positions of influence in the public and private sectors. Only then could real shifts in access to power and social mobility be measured at scale. Similarly, for an underserved family, it would take generations to climb the social ladder—own a home, find stable employment, access better living conditions, educate their children, and perhaps start a business.

Moreover, all of this had to happen under ideal conditions for change to be noticeable within one generation. But in reality, the process was far more complicated. Corruption among political leaders, economic downturns that led to high unemployment, regional tensions, mismanagement of immigration, and global crises like pandemics have all slowed the change process. The inertia of social systems and unpredictable external factors illustrate the complexity of equitable change.

In the context of Equitable and Inclusive Design, we need to recognize that the impact of our actions may take years, if not decades, to become measurable, especially when tackling systemic inequality. Unlike immediate product accessibility improvements, deeper changes in societal equality are gradual and may be difficult to observe in the short term.

In this framework, I advocate for defining nonmeasurable metrics—goals that are visionary and long-term. These goals might not be measurable during the project's lifecycle, but they provide direction. Typically they focus on social justice, equal opportunity, and eliminating disparities across society. For example, they might address goals like eradicating gender gaps, fostering economic equality, or creating a world where people of all abilities can fully thrive. These are goals that extend beyond the immediate scope of a project and might only come to fruition years down the road.

These metrics, although not intended to be directly measured, help set the tone for the project's narrative and clarify its long-term objectives. They guide discussions and give a clear sense of purpose, ensuring that the project is aligned with its ultimate vision. By doing so, they contribute to the success of the initiative by maintaining focus on the broader impact, helping teams stay motivated and united around a common mission.

Just as in South Africa's journey toward equity, the process of achieving systemic change through Inclusive Design requires patience, foresight, and a deep understanding that true impact unfolds over time.

Local vs. Global and the Replicability Puzzle

Another critical factor to consider is the replicability of your idea—how easily your solution can be transferred to different places, audiences, or points in time.

In an ideal scenario, the best solutions and practices would be easily transferable, achieving the same impact no matter where or when they were applied. However, the reality is more complex, and replicating an idea from one context to another can be challenging.

Sometimes, it's fairly clear why a solution would succeed in one specific context but struggle in another. However, there are many instances where the differences are more subtle, and the need for *adaptation* becomes apparent. I'll explore a few examples where solutions required adjustments or, in some cases, where replication simply wasn't feasible.

Focus on Mechanisms, Not Just Contexts

A key to the success of many Inclusive and Equitable Design projects lies in their ability to integrate cultural nuances that foster a sense of belonging for the target audience. For instance, when we redesigned a social media platform to support Black business owners, we embedded various cultural markers throughout the interface. We tailored elements like the choice of words, the emojis used, and the color scheme to create something that resonated with this community. What made the design so effective for this specific group, however, also meant that it wasn't universally applicable—the very elements that made it culturally relevant for one audience would make it less so for another.

Another example is the lucky iron fish, a product designed to combat anemia in Cambodia by releasing iron into food as it cooks. The product gained traction not just because of its functional benefits but because it was shaped like a fish—a symbol of good luck and prosperity in Cambodian culture. But this cultural significance doesn't translate universally, meaning the product would need to be adapted if introduced in other countries. A fish-shaped product might hold little meaning or relevance in places where the fish doesn't carry the same symbolism, which would make it less effective from a marketing and cultural perspective.

These examples illustrate that replicability in design is not just about copying a solution from one place to another but understanding the mechanism behind the success and adapting it to the new context. In the case of the social media platform, the mechanisms behind the design—such as community engagement, ease of navigation, and user accessibility—could easily be transferred to other platforms or audiences. However, the specific cultural signals embedded into the design, like wording and color schemes, would need to be adjusted for different audiences.

When considering the lucky iron fish, a similar product introduced in Mexico could take the form of corn, a plant considered sacred and symbolic of life and fertility in Mexican culture. Corn has deep cultural significance in the region, with roots in Maya mythology, where it is believed that humans were made from corn. Therefore, using a corn-shaped product would tap into a sense of cultural relevance and meaning similar to that of the fish symbol in Cambodia.

In Polynesia, the product might be designed to resemble a pig. In Polynesian cultures, pigs are symbols of wealth, success, and generosity. They play a key role in important cultural events and are often given as gifts during weddings, funerals, and other significant gatherings. By adjusting the product to reflect these cultural symbols, the design can maintain its practical purpose while becoming relevant and meaningful within the new cultural setting.

This same principle applies to the social media platform. Although the engagement mechanisms and user-friendly interface can remain consistent, the specific design elements would need to be customized to resonate with the new audience. For instance, when adapting the platform for Chinese business owners, we could incorporate red, a color that represents good fortune and prosperity in Chinese culture. We could also introduce ephemeral animations tied to significant events like Chinese New Year and adapt the content moderation policies to respect cultural differences. What may be seen as confrontational or aggressive in a Western context might be considered polite or even respectful in an Asian cultural setting, and vice versa.

At the heart of replicability is the understanding that although the underlying mechanisms—whether community engagement, product accessibility, or improved user interaction—can remain the same, the specific design features must be adapted to fit the local culture and context. Each time a solution is introduced to a new environment, it requires careful consideration of the cultural, social, and economic nuances that make the target audience unique.

Ultimately, it is the mechanisms behind the design—the process of community engagement, the focus on accessibility, and the ability to integrate seamlessly into daily life—that determine whether a solution can be replicated across different contexts. The specifics, however, must always be flexible enough to adjust to the local cultural landscape. This combination of adaptation and consistency is what allows successful designs to thrive in multiple settings while maintaining their original intent.

Adapt to Local Distribution Channels

When considering the distribution of a product, it's essential to think beyond just the product itself and consider how it will be delivered to the users.

For example, when we were working on making banking services more accessible to underbanked communities, we quickly realized that the communication channel was a significant issue. It wasn't just about literacy levels; it was about how the message was being delivered. Many users relied on simple phones using text messages or, in some cases, WhatsApp. These channels were part of their daily lives and represented spheres of trust. In contrast, a complex app developed by the bank felt alien and untrustworthy to them. To address this, we adapted the communication by shifting to text and WhatsApp for interactions. We used simple, clear messaging; visuals; and communication anchors that helped preserve their sense of trust in the system.

Similarly, when designing notebooks for people with neurodivergence, we also had to carefully consider distribution channels. In many areas, neurodivergence is still a taboo topic, and openly advertising a product for this audience could stigmatize it, leading to low sales and negative perceptions. In these cases, we opted to distribute through therapists and professionals who could introduce the product in a private, trusted setting. This approach allowed the product to reach its intended audience without the risk of exposure in a more public, potentially stigmatizing space.

By aligning distribution channels with user needs and cultural sensitivities, we were able to foster trust and adoption, ensuring that our solutions reached their target audiences effectively.

Scaling and Additionality

Another key consideration when scaling inclusive solutions is the economic context in which the solution is introduced. In some regions, the solution might be entirely new and fill a critical gap. However, when transferred to other contexts, it could inadvertently compete with existing solutions, potentially causing harm rather than adding value. This is where the concept of additionality comes into play.

Additionality refers to the process of determining whether a certain outcome or result is directly attributable to a specific intervention, action, or policy. It essentially asks, "Would this outcome have occurred without the intervention?" The idea is to assess the unique value that can be attributed to the intervention itself rather than to other unrelated factors.

When applying additionality to the scaling of inclusive and equitable solutions, it is essential to ensure that the new intervention doesn't cannibalize or disrupt a progressive solution that is already in place or in development. For example, introducing a new banking service to an underbanked community where local cooperative banking initiatives are already gaining traction could undermine the existing community-rooted solution. In such cases, it is crucial to assess whether the new solution is genuinely adding value or simply competing with something that is already effective.

By ensuring additionality, we not only protect existing solutions but also guarantee that the new intervention brings genuine progress to the community rather than merely shifting resources or focus from one solution to another.

The Nike Hijab Example

A notable example of additionality in the context of scaling is the Nike Pro Hijab. Nike's original intention behind this product was admirable: to allow women of all faiths, particularly Muslim women, to participate in sports comfortably while supporting their performance goals.

The Nike Pro Hijab was launched in December 2017 after years of development and feedback from Muslim female athletes. It was designed to tackle the specific challenges that traditional hijabs posed during athletic activities, such as discomfort, heaviness when wet, and limited hearing or movement. The lightweight, breathable fabric and pull-on design provided much-needed comfort and performance benefits. This innovation was a breakthrough for hijabi athletes like Ibtihaj Muhammad and Amna Al Haddad, both of whom contributed to the testing process.

Although the product made sense in the United States, where the values of sports align closely with Western ideals, its global expansion faced significant challenges. Cultural differences in how sports are perceived, as well as local nuances regarding modest wear, presented obstacles. For instance, in some regions, the presence of the Nike logo on religious apparel was seen as inappropriate or out of place.

Moreover, in many countries, local clothing makers were already offering hijabs designed for athletic use. The introduction of the Nike Pro Hijab posed a risk of displacing these local businesses, which was certainly not Nike's intention. In this case, Nike had to balance its noble goals with the potential negative impact on existing markets, illustrating the importance of considering local dynamics and additionality when scaling such innovations.

A Recipe for Scaling

Scaling any solution, whether it is an inclusive and Equitable Design or not, is always a significant challenge. However, the desire to scale is strong, particularly for designers and leaders who want to see their product or service widely implemented. This is especially true when the impact on users and communities goes beyond simple functionality and offers real societal benefits.

In my years of experience advising large organizations, I've realized that I've often learned more about scalability from them than the other way around. When I step back, it becomes clear that although large corporations may have many flaws and inefficiencies—especially from the perspective of hyperactive Silicon Valley entrepreneurs—no one can deny their ability to execute at scale. What these corporations accomplish in terms of processes, investments, planning, and skills to deliver their products and services on a massive scale is impressive, and it's important to recognize this.

With that in mind, I've been able to develop and suggest a simple recipe that can help increase the chances of success when scaling an equitable or inclusive design. Through my work with various organizations, I've identified five key steps or practices that can maximize the impact of a solution and improve its scalability.

This approach acknowledges the challenges of scaling while also leveraging the strengths and practices that large organizations have honed over time, blending them with the intentionality and vision required for equitable and inclusive solutions. These steps offer a framework for designers and leaders to ensure that their efforts not only reach more people but also do so in a meaningful and impactful way.

Prove the Success with Documented Metrics

Inclusion and equality are values that can be more difficult to quantify compared to purely technical or financial outcomes. As discussed earlier in relation to KPIs and impact measurement, it is crucial in every Inclusive Design project to document the success of the solution as early and as thoroughly as possible.

Just as with any product, metrics of success are essential to convince stakeholders to support and promote the solution to the next level. This is especially true for inclusive and equitable products, where the field can sometimes be seen as less concrete or harder to measure. Continuously and rigorously documenting the impact, successes, and failures of the project provides objective data to back up the decision to scale. DEI is often a sensitive and emotional subject, and decision-makers can be influenced by their feelings or fears, making hard data even more valuable.

For example, we worked with a blockchain startup at a time when blockchain technology was a very hot topic. As we developed the solution, the founder quickly recognized the need to document our progress and impact. He set a goal of publishing a research article to not only generate interest but also to legitimize the work we were doing. This became a great asset for the project. The team was able to build a compelling argument supported by solid data. When it came time to expand the solution, it was much harder for others to argue against a well-researched and thoroughly documented article than against a simplistic, unsubstantiated pitch.

In many contexts, it's said that women or people of color have to work extra hard to overcome stereotypes and systemic biases in the workplace. A similar concept applies to Equitable and Inclusive Design compared to more traditional innovation projects. Decision-makers care, but extra effort is often required to prove that your project is worth the investment. By documenting impact and success, you help build the case for your project's value—until a time when equality is fully achieved in these fields.

Attract Funds and Important Support to Grow

Just like with any project, scaling an equitable and inclusive solution requires investment—whether financial, human, or time-related. Equitable products are no exception, as they often necessitate changes in production, marketing, operations, or even organizational structures to ensure success.

To attract the necessary investment, it's critical to prove the business case behind the innovation. Referring back to the business cases discussed in Chapter 7, "Implementing Inclusive Solutions and the Business Cases Behind Them," there are numerous ways to demonstrate financial gains—or at the very least, ways to prevent losses. This is key when pitching to potential investors or securing internal support.

One challenge I've observed among designers and DEI advocates is that they don't always have the strongest business acumen. This isn't surprising, as those drawn to design and DEI careers often have backgrounds in the humanities and emphasize empathy, which can seem at odds with more financially driven approaches. However, these fields are far from incompatible. I've seen numerous cases where design teams managed to strengthen their business arguments and make compelling cases for investment.

A particularly interesting example comes from a project with a large consumer electronics company. The goal was to develop a smartwatch for people with visual impairments. The solution was elegant, built around three innovations: haptic braille technology (allowing braille words to be "felt" directly on the user's palm), a snap-on band for easier use and stability, and enhanced connectivity with the company's wider ecosystem for better navigation.

Despite its appeal, the project faced skepticism over its scalability and cost/benefit ratio. The team ultimately turned things around by presenting a clear market potential: millions of visually impaired people worldwide could benefit from an accessible, smart wearable device. But more importantly, they emphasized how no other product in the market provided the unique combination of their 360-degree solution and that this was a novel and unique approach to consumer electronics. They positioned their projects not only as a solution to an inclusion problem but also as a pioneering innovation project that was looking at electronic devices from three angles combined: interface, physical ergonomy, and ecosystem integration.

This pivotal moment helped the team gain crucial support from both external experts and internal executives, clearing the path for additional development and elevating the project's value within the company's innovation portfolio.

This story illustrates how a strong business case that combines market potential with social impact, combined with key internal and external support, can make the difference in scaling an Inclusive Design solution.

Work with Community Partners and Users to Build Credibility and Scale

Creating systemic solutions requires engaging with a broader system to drive change. Even the most well-intentioned and capable company cannot scale a solution in isolation. Success hinges on a company's ability to connect with its ecosystem, fostering collaboration to amplify the solution's reach and impact.

Your product doesn't exist in a vacuum—it's part of a network of stakeholders who can either support and enhance it or criticize and block its progress. One of the keys to scaling a solution is building an ecosystem around it, which brings together positive endorsements, complementary skills, and expanded execution capabilities.

A clear example of this is a project we did with an Asian pharmaceutical company looking to break through a growth ceiling in the U.S. market. The company's overarching goal was to regain the trust of minority communities, especially populations that had become wary of large pharmaceutical companies. We recognized that the company needed to connect with community players and local organizations to gain this trust. These organizations, already trusted by the target populations, would be invaluable allies in supporting the company's efforts.

We advised the company to engage with local associations, NGOs, and grassroots organizations—groups that were already in direct contact with these communities. Not only could these groups provide crucial feedback on the company's efforts, but their endorsements would help build credibility and bridge the trust gap.

This collaboration was a humbling experience for the pharmaceutical company, one of the world's largest multinationals. It had to approach small, volunteer-run groups and convince them to collaborate, which required adopting a new mindset. However, this step created a precedent and laid a solid foundation for the company's long-term scaling strategy.

By building partnerships with trusted community organizations, the company not only enhanced its distribution and credibility but also ensured that its solution had the support needed to grow sustainably.

The Speed Dilemma

When it comes to the industrial process behind inclusive and equitable solutions, it's tempting to remain consistent with the initial intention by selecting manufacturers and suppliers from underserved communities or diverse backgrounds. This can add significant co-benefits, such as creating jobs and development opportunities for those associated with the scale-up of the solution.

Although this is a noble and highly desirable approach, there is a caveat. Often, these suppliers may not possess the same ability to scale as more traditional, established players. This means their services could be more expensive, slower to deliver, or possibly of lower quality.

In some cases, where the timeline allows for a slower pace or where paying a premium and tolerating lower quality is feasible, working with these suppliers is the preferred choice. However, in situations where you're tackling an urgent problem, such as addressing a health crisis in a large population, speed becomes essential. The sooner a solution can be implemented and distributed, the faster the problem can be addressed, potentially saving lives or limiting the spread of an issue. In such contexts, there's often a dilemma: should you prioritize suppliers whose values align with your project, or compromise in favor of a faster, more efficient process to meet critical needs?

I've encountered this dilemma in several projects where we aimed to hire local manufacturers to support social and environmental goals: limiting cross-continent shipments and avoiding suppliers that rely on carbon-heavy processes. However, the trade-offs were challenging, and we often had to prioritize execution speed over our initial values, especially in the short term.

Timing and speed are often the biggest hurdles. Although quality, volume, and competitive pricing can usually be achieved if there's no time pressure, the interventions needed for a more inclusive and equitable society are usually urgent. Delays can widen gaps and create larger issues over time. A good example of this was the access to high-speed internet in low-resource communities across major U.S. cities. What initially seemed like a comfort issue became a major problem during the pandemic when the shift to online life exposed the digital divide. Millions lost their jobs, and children fell behind in school simply due to poor internet access. What seemed like a minor issue soon contributed to deepened inequalities. Families slipped into extreme poverty, reducing their chances of social mobility, and kids missing out on education faced long-term disadvantages in the job market.

When thinking about scaling, it's important to also consider speed. Delays in execution, especially for urgent matters, can have far-reaching consequences, compounding inequalities and creating more significant issues down the road.

Be Ready for the Unexpected

The final ingredient in the recipe for scaling is embracing the idea of being ready for the unexpected. As we've discussed, replicating a solution is never straightforward—what works in one context may fail or face challenges elsewhere.

Preparing for the unexpected may seem like an oxymoron, but it essentially means transforming uncertainty into manageable risks by anticipating potential failure points and adapting your infrastructure accordingly. This involves proactive risk management, where you forecast possible issues and build contingencies into your scaling strategy.

Some of this preparation comes down to planning. You can create a more resilient supply chain, diversify your partner and funding sources, and implement strategies to de-risk your business and operations. These measures ensure that you're not overly reliant on any single resource or partnership, which can help mitigate unexpected challenges.

Another key component is team diversity. As mentioned earlier, diversity is a powerful asset in building resilience. A diverse team brings a wider range of perspectives and skills, which is invaluable when facing unexpected challenges or adjusting to new markets. As your solution scales and reaches broader audiences, it's essential that your team reflects this growth, helping you stay adaptable and responsive to changing circumstances.

Finally, a lot of this readiness comes down to mindset. Scaling is a humbling process, requiring you to treat each new market or iteration as a fresh start. It's essential to approach every phase of growth with openness and humility, acknowledging that no single solution will work perfectly everywhere and that flexibility and adaptation are critical to long-term success.

Navigating the Challenges of Equity in a Changing World

When considering the implementation of Inclusive and Equitable Design—and, by extension, DEI practices—in professional environments, it is crucial to address several key factors.

Although the intentions behind these efforts are undeniably noble, their outcomes are often difficult to measure and demonstrate. This is partly because the true impact of such initiatives often requires a long-term, sometimes multigenerational perspective. For example, the effects of greater diversity on creativity, resilience, and team dynamics can be subtle and challenging to quantify. Furthermore, when businesses attempt to evaluate DEI's influence on high-level metrics like revenue or growth, a multitude of variables can interfere, obscuring the results. Additionally, success is rarely achieved on the first attempt, making patience and adaptability essential.

In light of these challenges, many organizations appear to be slowing down or even retreating from their DEI initiatives. Experts and critics are growing impatient or disillusioned with the tangible outcomes of many policies or pledges, and members of underrepresented communities are increasingly weary of being tokenized. This signals the need for an evolution in how we approach DEI—not abandoning the commitment to a more equitable society but refining the methods used to achieve it. The data is clear, and we discussed this in earlier chapters: inequality is rising, and the social divide continues to widen.[8, 9]

This is where Inclusive Design can complement and even replace traditional DEI practices. Unlike corporate training sessions or affinity groups, Inclusive Design focuses on creating practical, user-centered solutions that produce tangible, measurable results in both the short and long term. By prioritizing "curb cut" solutions—interventions that benefit everyone while addressing the needs of marginalized groups—Inclusive Design can reduce stigma and elevate the overall user experience. These approaches not only offer concrete benefits but also contribute to fostering a culture of equity and inclusion in a more organic and sustainable way.

That said, I recognize that Inclusive Design, like any approach, is not a silver bullet. It will likely encounter its own failures, and there may come a time when the enthusiasm surrounding it fades. However, I hope that some of its core principles will endure, becoming embedded as foundational elements within organizations. This kind of legacy, even if partial, can help sustain the momentum toward equity over time.

I also believe that new events and societal shifts will continually shape how we approach DEI. Historically, transformative moments often arise in response to opposing forces. For instance, the global outcry following George Floyd's tragic death galvanized a wave of activism and renewed focus on systemic racism. Similarly, the current challenges to democracy and the rise of far-right movements across the Western world may provoke new reactions, leading to fresh iterations of how we strive for equity.

As we navigate these evolving dynamics, the fight for a more equitable society will remain essential. The key is to adapt our strategies to meet the moment, staying committed to the ultimate goal while being flexible and innovative in our methods.

Notes

[1] ACLI, 2020. CEO action briefing. [Online]. Available at: `https://www.acli.com/-/media/acli/files/events/2020webinars/ceo_action_briefing.pdf` [Accessed: 20 January 2025].

[2] CEO Action for Diversity & Inclusion, n.d. CEO diversity pledge: all signatory list. [Online]. Available at: `https://www.ceoaction.com/media/4575/ceo-diversity-pledge-all-signatory-list.pdf` [Accessed: 20 January 2025].

[3] CEO Action for Diversity & Inclusion, 2018. CEO action FAQ. [Online]. Available at: `https://www.ceoaction.com/media/1794/ceo-action-faq-june-2018.pdf` [Accessed: 20 January 2025].

[4] Li, H., 2018. The application of TCM thought of preventive treatment of disease in clinical practice. *Advances in Social Science, Education and Humanities Research* 132, 91–95. [Online]. Available at: `https://www.atlantis-press.com/article/25889233.pdf` [Accessed: 20 January 2025].

[5] Stovall, J., 2018. How to get serious about diversity and inclusion in the workplace. TED. [Online]. Available at: `https://www.ted.com/talks/janet_stovall_how_to_get_serious_about_diversity_and_inclusion_in_the_workplace/transcript?subtitle=en` [Accessed: 20 January 2025].

[6] Bazelon, E., 2024. Harvard grapples with diversity in admissions after affirmative action ruling. *The New York Times*, 11 September. [Online]. Available at: `https://www.nytimes.com/2024/09/11/us/harvard-affirmative-action-diversity-admissions.html` [Accessed: 20 January 2025].

[7] HonestiValues, 2024. How can organizations promote diversity and inclusion in employee hiring and promotion practices? [Online]. Available at: `https://honestivalues.com/en/blogs/blog-how-can-organizations-promote-diversity-and-inclusion-in-employee-hiring-and-promotion-practices-56712` [Accessed: 20 January 2025].

[8] Horowitz, J. M., Igielnik, R., & Kochhar, R., 2020. Trends in income and wealth inequality. Pew Research Center. [Online]. Available at: `https://www.pewresearch.org/social-trends/2020/01/09/trends-in-income-and-wealth-inequality` [Accessed: 20 January 2025].

[9] Wike, R., Fagan, M., Huang, C., et al., 2025. Economic inequality seen as major challenge around the world. Pew Research Center. [Online]. Available at: `https://www.pewresearch.org/global/2025/01/09/economic-inequality-seen-as-major-challenge-around-the-world` [Accessed: 20 January 2025].

The Next Generation's Power: Rethinking Inclusion for a New Era

For years, diversity and inclusion efforts have been driven by corporate pledges, policy changes, and public commitments—but have they truly reshaped the systems they set out to change? As backlash grows and institutions struggle to maintain momentum, the next generation is stepping into a world where inclusion is no longer just a talking point but a battleground. In this chapter, we'll examine why diversity, equity, and inclusion (DEI) efforts have faced resistance, how young leaders and workers are uniquely positioned to push for lasting change, and why the future of inclusion depends on embedding it into the very fabric of business, technology, and society. This is not just about fixing the past—it's about designing a future where inclusion isn't an initiative but a given.

The DEI Reckoning: What Went Wrong?

The last few years have been a wild ride for DEI. What started as a movement filled with urgency and momentum—spurred by the protests of 2020 and the massive corporate pledges that followed—has quickly unraveled into something few could have predicted.

In 2025, President Donald Trump took office again, and one of the first things his administration did was draw a hard line against DEI efforts. The rhetoric was clear: only two genders, an end to corporate diversity mandates, and a rejection of what was labeled "woke capitalism."

A significant shift has occurred in the corporate landscape as several major companies have scaled back or entirely dismantled their DEI initiatives. This trend, which has swept through industries ranging from tech giants to automotive manufacturers, marks a notable departure from the previous emphasis on workplace diversity and inclusion.

Boeing, the aerospace giant, made headlines when it eliminated its global DEI department in late October 2024.[1] Under the leadership of new CEO Kelly Ortberg, the company integrated DEI staff into a broader human resources group focused on talent management and employee experience. This move was part of a larger restructuring effort that included a 10% reduction in overall headcount. Boeing's decision reflects a growing sentiment among some corporate leaders that DEI programs may not be delivering the intended results or aligning with new strategic priorities.

Other prominent companies have followed suit, each with its own approach to scaling back DEI efforts. Amazon, for instance, has removed DEI language from its website and stepped back from previous diversity goals.[2, 3] Google's parent company, Alphabet, no longer includes DEI references in its annual report, and Meta has disbanded its internal DEI team.[4] Even consumer-facing companies like McDonald's and Lowe's have made changes, renaming diversity teams and ceasing participation in external diversity surveys.[5, 6] These actions collectively signal a significant shift in how major corporations approach workplace diversity and inclusion.

It would be easy to look at this and say that DEI was a passing trend, that people never really wanted it, or that it was doomed from the start. But I don't think that's the full story. The backlash we're seeing today isn't necessarily about diversity, equity, and inclusion as principles—it's about how those principles were implemented. The problem wasn't the idea of fairness; it was the gap between the promises made and the actual, tangible benefits delivered to communities.

A Movement That Lost Its Way

Between 2020 and 2024, DEI became a booming industry. Every major company had a chief diversity officer. Entire consulting firms popped up offering workshops, playbooks, and corporate training sessions. There were LinkedIn posts filled with pledges, panel discussions with executives vowing to "do better," and diversity reports showing incremental progress in hiring. But behind the scenes, many of these initiatives were toothless, below are some feedback I heard many times from partners, clients or community organizations:

- ▪ Hiring targets were set, but retention rates stayed the same.

- Bias training became mandatory, but workplace culture remained unchanged.
- Funding was allocated to DEI, but actual salaries, opportunities, and promotions for marginalized groups barely moved.

For many people—especially those from underrepresented backgrounds—it started to feel like a massive corporate PR effort rather than a real transformation, or at least the effort felt very focused on the short term and what was visible. The system as a whole seems very resilient in the sense that after being shaken, it tends to revert to its original state. And if you're part of a community that has been historically excluded, seeing a company post a black square on Instagram doesn't change your paycheck or your lived experience.

At the same time, DEI also became a political battleground. While progressives pushed for more systemic change, conservatives increasingly framed DEI as ideological overreach, corporate virtue signaling, and, in some cases, even reverse discrimination. Universities became a focal point for these tensions, with accusations of DEI offices limiting intellectual diversity rather than expanding it. By 2025, DEI had become so charged that many companies and institutions decided it wasn't worth the fight anymore.

But the most interesting part of this backlash is who helped fuel it. Although many assume the anti-DEI wave is driven by white conservatives, a closer look at the 2024 election tells a more complex story.

The Real Story Behind the Backlash: Disillusionment, Not Rejection

The fact that many people of color—particularly Hispanic and Black voters—shifted toward Trump in 2024 wasn't just about cultural conservatism. It was about disillusionment. People weren't rejecting the principles of equity; they were rejecting the gap between rhetoric and results. They were tired of politicians and corporations talking about justice but failing to address the deeper issue: economic inequality.[7]

A working-class Hispanic voter in Texas or a Black small-business owner in Georgia wasn't necessarily saying, "I want less diversity in my workplace." They were saying, "How has any of this helped me feed my family, get healthcare, or buy a home?" And in many cases, they weren't wrong to ask.

The mistake of corporate DEI was assuming that symbolism was enough. That if companies did the trainings, changed their hiring language, and made the right statements, people would be satisfied. But symbolic efforts don't fix the structural issues at the heart of inequality. When you live in a world where wages are stagnant, the cost of living is skyrocketing, and the wealth gap is getting worse, no amount of corporate pledges will make you feel like things are improving.

The backlash we're seeing today is a reaction to that disconnect—not necessarily to the values of inclusion themselves, but to the way they were co-opted, diluted, and commodified by institutions that ultimately failed to deliver real change.

So, does this mean DEI is dead? No—but it does mean the next era of inclusion will have to look radically different. The days of performative activism and corporate messaging won't cut it anymore. If inclusion efforts are going to survive and thrive, they must shift away from ideological battles and focus on pragmatic, measurable outcomes that impact people's daily lives.

Rather than being framed as a political stance, inclusion needs to be approached as a practical necessity—something that delivers visible benefits rather than just symbolic gestures. Success should no longer be measured by the number of diversity training sessions completed or corporate pledges made but by real progress in economic mobility, access to opportunity, and the closing of systemic gaps.

For this to happen, inclusion cannot remain a separate, siloed initiative. It must be woven into the core of business models, embedded in product design, hiring practices, leadership development, and decision-making processes. If companies treat it as a side project, something that exists parallel to their real work, then it will always be at risk of being deprioritized or discarded when the political climate shifts. But if inclusion is built into the foundation of how organizations operate, it will be far more resilient—surviving not as a trend but as an essential component of long-term success.

The backlash to DEI is not the end of the conversation. It's an opportunity: an opportunity to rethink how we approach equity in a way that actually works. The next generation won't be won over by empty pledges—they will demand results. And those who can't deliver will be left behind.

The Next Generation and the Crisis They Will Inherit

The world young people are stepping into today is eerily similar to what happened after the post-World War II boom. The Baby Boomer generation lived through a period of massive growth and technological advancement, embracing a lifestyle that, at the time, seemed limitless. More consumption, more expansion, more extraction—without truly grasping the long-term consequences. The planet could take it, or so they thought. By the time the next generations realized the damage—climate change, environmental destruction, resource depletion—it was already too late to prevent a crisis.

The same is happening now, but with the very structure of our societies. For decades, we've been dismantling systems of support, solidarity, and economic fairness, often without acknowledging it. Public debt has skyrocketed, and wealth has been increasingly concentrated in the hands of a few. Regulations that once protected the working and middle classes have been eroded, leaving entire communities vulnerable to market forces that only serve those already at the top. The result is an extreme imbalance—one that the next generation will be forced to navigate.

An example of this can be seen in the declining rate of unionized workers in the United States, which has reached an all-time low. In 2024, only 9.9% of wage and salary workers were union members, a slight decrease from 10.0% in 2023. This trend raises concerns about the implications for both businesses and employees.[8]

Proponents of a less unionized workforce argue that it offers businesses greater flexibility, potentially fostering economic growth and innovation. However, the diminishing presence of unions can leave employees more isolated and with reduced bargaining power, impacting their ability to negotiate for better wages and working conditions.

Research indicates a notable correlation between declining union membership and increasing income inequality (see Figure 9.1). As union density decreases, a larger share of national income tends to concentrate among the wealthiest segments of society. This relationship suggests that robust union representation has historically played a role in promoting a more equitable distribution of income.[9]

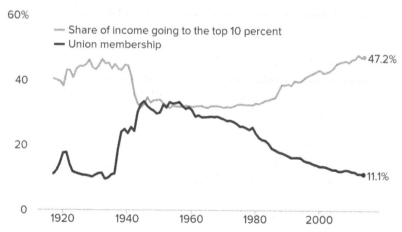

Figure 9.1: Union membership and share of income going to the top 10%, 1917–2017 (Economic Policy Institute)

The ongoing reduction in unionized workers not only affects individual employees but also has broader societal implications, potentially exacerbating wealth disparities and altering the balance of economic power.

What makes this crisis even more insidious is that it has been hiding in plain sight. Although progress on social issues has been widely celebrated, the economic structures that shape opportunity, mobility, and stability have remained largely untouched. We talk about equity, but we exist in a world where the gap between the richest and the poorest has never been wider, where billionaires accumulate wealth at an unfathomable pace while millions struggle with rising costs of living. DEI efforts over the last decade have been important in shifting perspectives and broadening conversations, but they have barely scratched the surface of these deeper structural issues.

This is the reality that the next generation will inherit—not just an ideological debate about diversity, but a society where foundational inequalities have been allowed to grow unchecked. They will wake up to the consequences of policies and decisions that prioritized short-term gains over long-term fairness, where the notion of inclusion wasn't dismantled outright but absorbed into a system that ultimately reinforced the same hierarchies.

Yet if any generation is equipped to push back, it's this one. They have unprecedented access to information and the ability to mobilize on a global scale. They can see through corporate messaging faster than any before them, and they understand their power as both consumers and talent. At the same time, they risk becoming a generation that is highly vocal but not action-driven—one that calls for change but remains trapped in cycles of outrage, engagement, and inaction.

What they lack is not awareness but leverage. They are entering a world where institutions are more fragile than they appear, where the companies that shape industries are desperate for young, talented minds to keep them relevant. Unlike past generations, they don't just inherit the economy—they *are* the economy. Their choices, their expectations, and their demands will shape the next era of inclusion, not as a social ideal but as an economic reality.

The question is whether they will use this influence to push for meaningful, long-term change—or whether they will allow the system, as it always does, to bend just enough to look different while staying exactly the same.

Inclusive Design as the Next Phase of Real Change

If the last decade has shown us anything, it's that performative gestures alone won't shift the balance. Inclusion cannot survive as a series of corporate statements, policy documents, or training sessions. It has to be something people can feel in their everyday lives—something that changes how they work, how they interact with products and services, and how they access opportunities.

This is where Inclusive Design has the potential to succeed where traditional DEI efforts have struggled. Unlike corporate DEI initiatives, which often operate as separate programs, Inclusive Design is not an external add-on: it is embedded directly into how companies build, create, and serve. It is not about fulfilling quotas or compliance; it is about designing systems that naturally accommodate and uplift diverse users from the start.

We've already seen how companies that integrate inclusion into their core operations can drive both social impact and economic success. Take the example of financial technology. In recent years, digital banking services have expanded access to financial tools for populations that traditional banks have historically excluded. Platforms like Cash App and Chime have intentionally designed their products to reach the underbanked, offering features that remove barriers such as minimum balance requirements and hidden fees. The result is a financial system that, by design, includes those who were previously left out—not as a special initiative but as a fundamental business strategy.

I've seen this shift firsthand in my work with companies that initially approached DEI as a compliance-driven exercise—something to report on rather than something that could drive value. One executive I worked with in a major real estate firm admitted that for years, the company's DEI efforts had been little more than checkboxes and internal training modules. It wasn't until the firm started incorporating Inclusive Design into its product development—rethinking the way it designed residential spaces to be more accessible, more adaptable, and more attuned to diverse community needs—that it began to see real business returns. The company's new projects didn't just meet compliance requirements; they attracted a broader range of buyers, strengthened community trust, and even improved long-term property values. That was the moment the leadership team realized that inclusion wasn't just a moral stance—it was a strategic advantage.

The same applies in the world of product design. In the tech industry, accessibility features that were once seen as niche considerations are now standard because they benefit everyone. Apple's VoiceOver, originally developed for visually impaired users, is now widely used by millions, from people with dyslexia to those who simply prefer a hands-free experience. What started as an inclusive feature became a universal tool—proof that designing for the margins can create value for the mainstream.

This shift toward built-in inclusion is not just good ethics; it's good business. Companies that ignore Inclusive Design risk alienating massive segments of the population—segments that are growing in economic and social influence. By 2045, the United States will no longer have a single racial or ethnic majority. The next generation of consumers and workers expects inclusion not as an added benefit but as a baseline requirement. Businesses that fail to adapt to this reality won't just fall behind on social responsibility—they will lose market relevance altogether.

What makes Inclusive Design powerful is that it doesn't rely on cultural tides or political trends. It operates at the level of function, shaping the everyday experiences of people in ways that can't easily be undone. Corporate diversity programs may come and go depending on leadership changes or political shifts, but Inclusive Design—when done right—creates systems that are hard to reverse because they work better for everyone.

The Danger of Monolithic Thinking: Teaching People to Think Independently

One of the biggest challenges in today's conversations about inclusion isn't just resistance: it's rigidity. In many spaces, DEI has come to be seen as a set of unquestionable truths rather than a constantly evolving process. The same institutions that advocate for open-mindedness sometimes struggle to tolerate genuine debate, dismissing dissenting opinions too quickly rather than engaging with them. The result is an environment where people feel pressured to either conform or stay silent, even when they have real concerns or alternative perspectives.

This is a problem not just for inclusion efforts but for society as a whole. True equity isn't about enforcing a single worldview; it's about making space for multiple perspectives while maintaining a commitment to fairness. The goal should be not to tell people what to think but to equip them with the tools to think critically.

This is exactly what I try to do in my classroom. I want my students to be challenged, to have their assumptions questioned, and to learn to navigate complexity rather than settle for easy answers. One of the best ways to do that is to bring in voices they don't expect to hear.

I once invited a pro-plastics lobbyist to speak in my Deplastify the Planet class. As you can imagine, this wasn't exactly a popular move. Students had spent the semester working on ways to reduce plastic waste and create more sustainable systems, and suddenly they were face to face with someone arguing that plastics were not the enemy, that banning them could have unintended consequences, and that the industry had a legitimate role to play in global development. The initial reaction was frustration, even anger.

But something interesting happened when they engaged. Some started pushing back with evidence and counterarguments. Others realized there were nuances they hadn't considered—like how plastic enables access to clean water and medical supplies in parts of the world where alternatives don't yet exist. The conversation didn't change their mission, but it made them sharper, more prepared to defend their ideas, and more aware of the complexity of the issue.

This is what true education should do. It should make students uncomfortable in the best possible way—not to force them to change their minds, but to ensure that they actually know why they believe what they believe. Too often, we mistake intellectual alignment for progress, assuming that agreement is the goal. But real progress comes from rigorous, open-ended questioning—from recognizing that the world is filled with contradictions and trade-offs that don't fit neatly into ideological boxes.

Take a debate that I like to spark in my class among students: blockchain technology. Depending on who you ask, it is either a revolutionary force for financial freedom or an overhyped, environmentally destructive scam. Both perspectives contain elements of truth. Blockchain has enabled decentralized finance and empowered individuals in countries with unstable banking systems, but it has also led to speculative bubbles and massive energy consumption. The reality is complex, and anyone engaging with the topic needs to weigh both the opportunities and the risks. Not once has this debate been sterile or one-sided; students always thank me and their peers for broadening their perspectives and offering alternative visions that ultimately enrich and strengthen everyone's arguments.

The same nuanced thinking should apply to inclusion. It's not enough to say that diversity is good—we need to ask when, where, and how it leads to better outcomes. We need to be willing to examine what's working and what's not without fearing that doing so will weaken the cause. If DEI efforts have failed in certain areas, it's not because inclusion is the problem: it's because the approach may have been flawed. The only way to improve is to acknowledge that reality and adapt.

This is particularly relevant in higher education. Universities are meant to be places of exploration and critical discourse, yet they have increasingly become battlegrounds where certain viewpoints are considered unacceptable. When ideological purity replaces intellectual diversity, we don't get progress—we get stagnation. Inclusion efforts should be about expanding conversations, not narrowing them. That means being open to difficult discussions, even when they challenge the dominant narrative.

The next generation will inherit a world that is more complex, polarized, and uncertain than ever before. They will need to navigate competing realities, question entrenched beliefs, and find solutions that work in practice, not just in theory. If they are only taught what to think, they will be ill-equipped for that challenge. If they are taught *how* to think, they will be the ones who redefine what inclusion looks like in the decades to come.

The Next Generation's Awakening: Recognizing Their Influence in the Corporate World

One of the biggest misconceptions young people have when entering the workforce is that they hold little power. They see companies as massive, impenetrable entities where they are just another name in the hiring process, lucky to land a job, and even luckier to have a say in how things operate. But that's not the reality—especially not today.

If there is one thing I always emphasize to my students, it's that companies need them more than they realize. There is a war for talent happening across industries, and businesses—especially the ones that want to remain competitive—are desperate to attract and retain young, highly skilled employees. And yet most students don't see themselves as having leverage. They spend months polishing their résumés, hoping to get through a brutal hiring process, all while assuming that they are the ones being evaluated.

But when I talk to companies, I hear a different story. Executives often express frustration that they struggle to understand and appeal to Gen Z and young millennial talent. They know that young employees bring fresh perspectives, digital fluency, and new expectations about workplace culture. They also know that if they can't evolve to meet those expectations, they will lose the best talent to competitors that do.

This is where inclusion takes on a new dimension. DEI has traditionally been framed as a corporate responsibility: something businesses should do because it's the right thing. But for young workers, inclusion isn't just an ethical issue; it's a baseline requirement. They expect workplaces to be diverse, equitable, and flexible—not because it's politically correct, but because that's the world they have grown up in. They want companies to reflect their values, not just pay lip service to them.

Looking at the composition of my classes at UC Berkeley and MIT, I can confidently say that diversity is a reality. However, I'm fully aware that these classrooms may be more the exception than the norm. These are highly progressive and open-minded environments, and on top of that, the students who enroll in my classes are often naturally drawn to the concepts of diversity and inclusion—some even actively fighting for their rights and striving for greater recognition of their differences. Yet whether in terms of gender, ethnicity, sexual orientation, age, socioeconomic background, or other dimensions, the groups I've taught have always been remarkably diverse and, just as importantly, deeply respectful of one another.

For these young talents and future leaders, this atmosphere and these standards are simply a given. They have grown up in environments where inclusivity is expected, not debated. The real question is whether they realize how exceptional this is—and whether they are prepared to carry these values into spaces that are far less accepting.

We've already seen how this plays out. Look at the rise of employee activism in major companies. Workers at places like Apple and Amazon have organized to push back against workplace discrimination, unfair labor practices, and ethical concerns about company policies.

At Apple, the #AppleToo movement emerged in 2021 as employees began sharing personal accounts of harassment, discrimination, and workplace mistreatment. It started as an internal Slack channel and later moved to a public website, where workers anonymously posted their experiences. Reports included allegations of gender and racial bias, retaliation for speaking out, and a general culture of secrecy that discouraged accountability. Within days, the movement collected hundreds of testimonies, exposing systemic issues that Apple's leadership had long ignored. The company was eventually forced to respond, leading to internal reviews and increased scrutiny over its workplace policies.[10]

A similar wave of activism hit Amazon, where warehouse workers organized to protest unsafe working conditions and demand better wages. One of the most significant moments came in 2022, when Amazon employees in Staten Island successfully formed the company's first-ever U.S. labor union—a move that shocked many, given Amazon's aggressive anti-union stance. Their victory signaled a shift in worker power, showing that even in corporate giants known for strict internal controls, employees could organize and push for change.[11]

These movements reveal a critical transformation in workplace culture. Employees—particularly younger generations—are no longer content with just having a job; they expect fair treatment, transparency, and real accountability from the companies they work for. And when those expectations aren't met, they're willing to organize, go public, and fight back.

I remember a conversation with a former student who had just landed a job at a major tech company. She told me that during her first few months, she felt like she had to stay quiet, to prove herself before questioning anything. But then, in a product meeting, she spoke up about the lack of accessibility features in an upcoming launch. To her surprise, leadership not only listened but actively asked for her input. She realized that the company wasn't just hiring young talent for their technical skills; it was looking to them for insight into what the next generation values, how they think, and what they expect from the world. The moment young employees stop thinking of themselves as passive players and recognize that companies are adapting to them, they can start leveraging that power to create the workplaces—and the future—they actually want to be part of.

Beyond workplace culture, young people's expectations are shaping consumer markets as well. The success of brands like Patagonia, which actively aligns with environmental and social justice causes, is a clear example of how businesses that take a stand can attract loyal customers. Meanwhile, companies seen as out of touch or hypocritical—those that make diversity pledges but fail to back them up with real action—quickly find themselves called out, losing credibility and market appeal.

The real shift that needs to happen is for young workers to recognize this power and use it strategically. Right now, many of them are aware of their influence, but they tend to express it through rejection—boycotting companies, calling out brands, or quitting jobs that don't align with their values. But real change doesn't just come from rejection; it comes from transformation. Walking away is sometimes necessary, but staying and demanding better can be even more powerful.

This is the conversation I push my students to have with themselves. How do you engage with a system that you want to change? Do you refuse to participate altogether, or do you step inside and push from within? There's no single right answer, but what's clear is that ignoring their own influence is a mistake.

Unlike previous generations, they don't just inherit the economy—they define it. The workplaces they choose, the businesses they support, and the expectations they set will determine whether inclusion becomes a passing trend or a fundamental part of how industries operate. If they walk into companies believing they have no power, they'll be treated that way. If they walk in knowing that companies are competing for their talent and values, they can start shaping the future from the inside.

From Performative Change to Embedded Inclusion: The Next Chapter

The backlash against DEI, the disillusionment of workers and voters, the resilience of exclusionary systems—these are not signs that inclusion has failed. They are signals that the way we have pursued it needs to evolve. The last decade has shown us what doesn't work: corporate pledges without action, diversity efforts siloed from real decision-making, and symbolic changes that don't alter fundamental inequalities. But the demand for fairness, representation, and equal opportunity hasn't disappeared. If anything, it's stronger than ever.

The next generation is stepping into a world that is deeply divided, structurally resistant to change, and increasingly shaped by corporate and technological power. They have the potential to redefine what inclusion looks like—not through policies that can be undone with a change in leadership, but through systems, products, and businesses designed for equity from the start.

This shift will require a different mindset and a different set of tools. It means moving away from compliance-driven DEI and toward inclusive innovation. It means seeing diversity not as an HR metric but as a strategic advantage in business, technology, and governance. It means designing institutions that naturally foster inclusion rather than trying to retrofit it onto structures built for exclusion.

Notes

[1] Aerospace Global News, 2024. Boeing dismantles diversity department. Aerospace Global News. [Online]. Available at: `https://aerospace globalnews.com/news/boeing-dismantles-diversity-department`.

[2] Heaton, R., 2025. What companies are rolling back DEI policies in 2025? TechTarget. Available at: `https://www.techtarget.com/whatis/feature/What-companies-are-rolling-back-DEI-policies`.

[3] Davis, D.-M., 2025. Amazon deletes "inclusion and diversity" language in latest filing. TechCrunch. [Online]. Available at: https://techcrunch.com/2025/02/07/amazon-deletes-inclusion-and-diversity-language-in-latest-filing.

[4] Fischer, S. & Allen, M., 2025. Meta rolls back DEI programs in latest bow to Trump. Axios. [Online]. Available at: https://www.axios.com/2025/01/10/meta-dei-programs-employees-trump.

[5] Arnold, A., 2025. McDonald's scales back DEI initiatives, halting diversity efforts. Finance Monthly. [Online]. Available at: https://www.finance-monthly.com/2025/01/mcdonalds-scales-back-dei-initiatives-halting-diversity-efforts.

[6] Parisi, K., 2025. McDonald's rebrands its DE&I team, but says it's "committed to inclusion," after threats from Robby Starbuck. HR Brew. [Online]. Available at: https://www.hr-brew.com/stories/2025/01/08/mcdonald-s-rebrands-its-de-and-i-team-but-says-it-s-committed-to-inclusion-after-threats-from-robby-starbuck.

[7] Brown, M., Figueroa, T., Fingerhut, H., & Sanders, L., 2024. Young Black and Latino men say they chose Trump because of the economy and jobs. AP News. [Online]. Available at: https://apnews.com/article/9184ca85b1651f06fd555ab2df7982b5.

[8] Durbin, D.-A., 2024. US unions flexed their muscles last year, but membership rates fall to record low. AP News. [Online]. Available at: https://apnews.com/article/unions-membership-rates-uaw-government-a3fc7bc50dd59a89f414230e8837d7e6.

[9] Sirota, D., 2021. The Labor Day graph that says it all. Jacobin. [Online]. Available at: https://jacobin.com/2021/09/labor-day-chart-union-membership-share-top-10-percent-income-inequality.

[10] Schiffer, Z., 2021. Apple employees are organizing, now under the banner #AppleToo. The Verge. [Online]. Available at: https://www.theverge.com/2021/8/23/22638150/apple-appletoo-employee-harassment-discord.

[11] Purifoy, P., 2025. Amazon Staten Island warehouse is retailer's first to unionize. Bloomberg Law. [Online]. Available at: https://news.bloomberglaw.com/daily-labor-report/amazon-staten-island-warehouse-is-retailers-first-to-unionize.

The Future of Inclusive Design

As I write these pages, I can't help but feel that our society is at a critical cross-roads. Everywhere I look, there is deepening polarization. People are losing faith in democracy, and the gap between the wealthy and everyone else has grown to levels that feel unsustainable. The signs of unrest and frustration are everywhere, and it often feels like the solutions are elusive. The journey toward inclusiveness and equality feels long and difficult, full of unexpected challenges. Yet, even in the face of this uncertainty, I hold on to hope. It's not blind optimism but rather a belief in our ability to adapt and create meaningful change. Through my work, I try to inspire others to share in this hope to recognize that progress is possible even when the road ahead seems unclear.

The future of Inclusive Design is not a distant abstraction; it is something we actively shape with every decision we make today. At a time when society is reckoning with systemic inequities, technological disruption, and cultural polarization, the way we design products, services, and policies will determine whether we move toward greater inclusion or reinforce existing divides. Although the challenges are significant, the possibilities for transformation are equally vast.

Designing for equity requires not just an understanding of present conditions but also a commitment to imagining what could be. This chapter explores how we can engage with the future in a way that is both rigorous and creative. Through the lens of design fiction, we will examine how speculative thinking allows us to break free from ingrained assumptions and envision futures that center on inclusion. We will also explore Distributivity and Additionality as essential frameworks that can guide us toward more equitable outcomes, ensuring that innovation contributes to collective progress rather than concentrating benefits among the privileged few.

This chapter is an invitation to think expansively and strategically. It is a call to reject the passive acceptance of the future as something that simply unfolds and instead recognize our collective power to shape it with intention. Through a combination of theory, practical tools, and scenario-based exploration, we will chart a path toward a future where Inclusive Design is not just an aspiration but a lived reality.

Future Modeling and the Decolonization of Futures

Future Modeling is one of my core areas of expertise: an evolution of Design Fiction that expands how we explore possible futures. Instead of predicting a singular, inevitable trajectory, Future Modeling maps out multiple possibilities, challenging assumptions and expanding the realm of what is conceivable. It enables us to navigate complexity, embrace uncertainty, and build long-term, systemic approaches to inclusion and equity.

The idea that the future is not predetermined but actively constructed through our choices, policies, technologies, and cultural narratives is central to this approach. However, many dominant narratives about the future are shaped by corporate interests, political agendas, and Western-centric visions of progress. The ability to imagine and share different futures—ones that break away from these entrenched structures—is essential if we want to cultivate a world that is genuinely more inclusive and equitable. Engaging in speculative futures that center marginalized voices and challenge dominant paradigms allows us to create a space for alternative narratives: ones that prioritize solidarity, equity, and systemic justice.

This idea aligns with Roman Krznaric's book *The Good Ancestor* (WH Allen, 2020), which argues that we must move beyond short-term thinking and instead act as stewards for future generations. The book explores how societies, businesses, and governments often prioritize immediate gains over the well-being of future citizens, reinforcing inequalities that persist across time. To counter this, Krznaric advocates for adopting long-term-thinking models, expanding our moral and political concerns beyond the present moment. Future Modeling provides a practical framework for embedding Inclusive Design into future-oriented strategies, ensuring that we build systems that stand the test of time rather than reacting only to short-term crises.

Decolonizing the Future

Decolonizing the future means actively questioning who gets to define what progress looks like. Historically, dominant narratives about the future have been shaped by Western economies, corporate giants, and political institutions that prioritize profit, control, and efficiency over inclusivity, sustainability, and justice. These narratives present technological innovation as inevitable progress while sidelining alternative visions that emphasize community-led governance, ecological balance, and localized economic models.

Consider, for example, how automation and AI are often presented as the future of work. The dominant discourse focuses on efficiency, cost reduction, and scaling productivity—prioritizing corporate interests while largely ignoring the displacement of low-income workers, the impact on local economies, and the cultural loss of human-driven industries. Meanwhile, alternative narratives of the future—ones that focus on cooperative work models, distributed ownership of technology, and AI as a tool for human empowerment rather than replacement—are rarely given the same level of attention.

If we are to truly decolonize the future, we must first liberate our imagination. This requires deliberate efforts to integrate diverse worldviews into the process of future-building. Indigenous knowledge systems, grassroots movements, and historically marginalized communities offer rich, alternative perspectives on progress that challenge the extractive, consumption-driven models that dominate today's global landscape. By embedding these perspectives into Future Modeling, we create space for futures that do not replicate the inequalities of the present but instead lay the groundwork for a radically different, more just society.

Imagine a future city built not around the logic of corporate urbanization but around principles of community-led infrastructure, where housing, transportation, and energy grids are cooperatively owned and governed by residents rather than large real estate developers. Imagine a digital economy that prioritizes shared ownership, where platform workers become co-owners of the technology they use daily. These alternative futures exist within the realm of possibility, but they require active engagement, vision, and advocacy to bring them to life.

The Role of Design Fiction in Future Modeling

Design Fiction, a term coined by Bruce Sterling and expanded on by Julian Bleecker, is a method of exploring potential futures through speculative storytelling and critical design. Instead of treating the future as a linear extension of the present, Design Fiction invites us to consider multiple, divergent possibilities—some plausible, others provocatively disruptive. It forces us to challenge our assumptions about what is inevitable and allows us to imagine alternative pathways that could emerge.

At its core, my practice of Design Fiction is about uncovering the futures that remain invisible to us due to our education, professions, responsibilities, and incentives. Strategic decision-making often relies on a narrow set of data points, filtering complexity into simplified trends. Although this makes information digestible, it also erases nuance, blind spots, and radical possibilities. For me, Design Fiction acts as a counterbalance, offering scenarios that go beyond the comfortable median and instead explore the disruptive, the unconventional, and the transformative.

Future Modeling draws inspiration from Design Fiction while adding key steps that transform reflections into a structured model of possible futures, identifying the switches that could enable them. It is also highly adaptive to the specific context of its audience and objectives, creating highly tailored scenarios—in this case, exploring what inclusive and equitable futures could look like.

Although Design Fiction often focuses on speculative storytelling, Future Modeling integrates those narratives into strategic frameworks that organizations and societies can use to remain adaptable and proactive in shaping an inclusive future.

Why Future Modeling Matters for Inclusion

One of the fundamental reasons we need Future Modeling is that we cannot predict the future. We do not know what the next George Floyd moment will be, what the next radical political shift will bring, or which technological disruption will completely reshape our world. This uncertainty is not a reason for passivity; it is a call to action. Rather than relying on predictions, we must adopt a mindset of continuous observation and preparedness. This means staying alert to societal shifts, weak signals, and emerging movements, and committing to long-term, systemic, and deeply rooted efforts like Inclusive Design that can withstand unpredictability.

Future Modeling helps us keep an open mind, allowing us to explore the full spectrum of what might be possible rather than being caught off guard by unexpected disruptions. By investing in resilient, future-proof strategies—ones that prioritize equity, accessibility, and adaptability—we can ensure that no matter how history unfolds, our commitment to inclusion remains steadfast and effective.

For example, think about how social justice movements gain traction. The Black Lives Matter movement was not a sudden phenomenon: it was a response to decades of systemic injustice, enabled by digital tools, community organizing, and a shifting cultural awareness. Future Modeling urges us to look at similar patterns emerging today—in labor rights, environmental justice, and the ethics of AI—to anticipate where major shifts might occur and how we can prepare to support inclusion within them.

In the next section, we will use Future Modeling to present five possible futures for Inclusive Design. Each scenario will explore different trajectories—some promising, others cautionary—allowing us to understand what is at stake and what forces might shape the world ahead. These scenarios will not be predictions; rather, they will serve as a tool to inspire critical thinking, proactive decision-making, and a more equitable approach to innovation.

Exploring Future Scenarios for Inclusive Design

The future is an open field of possibilities, shaped by global movements, technological shifts, environmental crises, and cultural transformations. The scenarios outlined here are just a small selection of potential futures; there could easily be hundreds more, each with different implications for society, equity, and inclusion. Some of these scenarios are inspired by work I did with School lab in 2024–2025 under a project called Future By Design, where we explored speculative futures and their implications for businesses, institutions, and communities.

Each of these five scenarios presents a high-level vision of how inclusivity and equity might evolve, followed by a narrative that brings the world to life, a set of key switches—external environmental factors that could push us toward this reality—and a critical question that forces us to reflect on its consequences.

These scenarios are neither utopian nor dystopian, nor do they represent my personal vision for the future. Instead, they serve as a tool to shift perspectives, prompting us to critically examine what kind of future we want to create and the questions we need to ask to get there.

The Commons Economy: Cooperative Ownership as the New Standard

In this future, global movements successfully push for community-owned platforms and services that challenge corporate monopolies. Worker-owned digital platforms, decentralized infrastructure, and cooperative AI models replace extractive capitalism. Economic power is redistributed as more people take ownership of the digital and physical systems that sustain their communities.

A Day in This Future

You call a ride-hailing service—not through a tech giant but through a community-owned mobility network, where drivers are stakeholders rather than exploited gig workers. Your home's electricity is managed by a local solar grid cooperative, and excess energy is traded among neighbors instead of being funneled to major corporations. In major cities, digital infrastructures are managed by cooperatively governed AI systems, eliminating the need for intermediaries and shifting power away from centralized tech firms.

What started as small grassroots initiatives—workers forming platform co-ops, communities building self-sustaining infrastructure—expanded into an economic transformation. Governments, pressured by citizen movements, created legal pathways to support community ownership models. As corporate monopolies failed to respond to rising inequality, people sought alternative systems that offered stability, transparency, and self-determination.

Switches That Enable This Future

- Severe economic downturns and job automation force communities to seek self-sustaining models of ownership.
- Climate-related energy shortages lead to the widespread adoption of localized, cooperative-owned power grids.

Critical Question

Can cooperative ownership scale without being co-opted by existing power structures?

AI as the Ethical Guardian: Bias-Free Decision-Making or Algorithmic Control?

In this future, AI serves as a global fairness regulator, eliminating human bias in hiring, law enforcement, and resource distribution. No decision affecting an individual is made without an AI ethics audit, ensuring that race, gender, and socioeconomic background do not influence outcomes.

A Day in This Future

A hiring manager at a major company no longer has the power to unconsciously favor one candidate over another. Every job application is blindly processed by AI, ensuring that only skills and experience matter. Police officers no longer decide whom to stop and search; an AI-driven oversight system prevents racial profiling in real time. Court rulings are checked for bias before being finalized, ensuring that justice is data-driven rather than emotion-driven.

What started as a movement to correct biases in AI systems transformed into a broader effort to replace human-led decision-making with algorithmic fairness checks. Activists fought for transparency in AI governance, but as more flawed human decisions came to light, society began trusting AI more than people for fairness. However, some worry that we've outsourced morality—if an AI tells us what is fair, do we still understand fairness ourselves?

Switches That Enable This Future

- A large-scale corruption scandal or judicial failure forces governments to implement AI-driven fairness audits.
- Extreme political polarization makes human-led governance unworkable, leading to a demand for algorithmic oversight.

Critical Question

Can AI enforce fairness without imposing a new form of technocratic control?

Neo-Tribalism: The Fragmentation of Society into Micro-Identities

In this future, society fractures into highly personalized microcommunities online and in the physical world. People increasingly live in ideologically or culturally homogeneous spaces, finding comfort in hyper-curated environments while avoiding interactions with those who think differently.

A Day in This Future

Your social media feed, entertainment choices, and even local businesses cater exclusively to your beliefs, values, and lifestyle. Neighborhoods are designed around shared identities, from vegan-only districts to AI-free enclaves where digital technology is banned. News media no longer attempt to present balanced viewpoints—every group has its own version of reality.

Hyper-personalization, AI-driven content ecosystems, and ideological reinforcement loops made this possible. As mass media and national identity structures broke down, people retreated into communities where they felt seen, heard, and unchallenged. Although many feel safer in these environments, society at large has lost the ability to communicate across differences.

Switches That Enable This Future

- A major misinformation crisis breaks public trust in centralized media, driving people into closed, niche information networks.
- Rising mental health concerns due to information overload push people to seek simpler, curated community experiences.

Critical Question

Does hyper-personalization destroy social cohesion, or does it create spaces where people thrive?

The Post-Diversity Workplace: Hiring for Adaptability, Not Identity

In this future, companies no longer track diversity through demographic quotas or representation metrics. The idea of hiring based on identity markers has been replaced by a new paradigm: selection based on adaptability, intersectional thinking, and resilience. Diversity is no longer measured in statistics; instead, it is embedded in how organizations value lived experience, complex problem-solving, and the ability to navigate an ever-changing world. The focus is not on who you are but on how you engage with difference, power, and uncertainty.

A Day in This Future

A job interview no longer starts with a resume. Instead, candidates share their lived experiences: what they've overcome, how they've adapted, and the unique perspectives they bring. A woman who raised three children alone is recognized for her crisis management skills. A refugee who navigated multiple languages and cultures is valued for their ability to bridge perspectives. A person who grew up without stable housing is seen as an expert in resilience and problem-solving under pressure.

What began as a response to labor shortages and the collapse of traditional education became a redefinition of merit itself. Companies stopped focusing on demographic quotas and started valuing the depth of human experience as an asset. Some hail this as the most meaningful form of inclusion—finally recognizing skills that were invisible for too long. Others question if abandoning traditional diversity metrics means erasing the struggles that made them necessary. If we stop measuring identity, are we truly beyond discrimination, or are we just better at ignoring it?

Switches That Enable This Future

- The decline of universities as gatekeepers of knowledge leads businesses to develop new, experience-based assessment models.
- A global talent shortage makes rigid hiring filters obsolete, forcing companies to value adaptability over credentials.
- Geopolitical shifts and systemic crises make intersectional problem-solving the most valuable skill, rather than fixed technical expertise.

Critical Question

Does removing identity from hiring create a more inclusive workforce, or does it erase the need to address systemic inequality?

The Resilience Index: Measuring Inclusion Like We Measure Carbon

In this future, inclusivity is quantified and tracked much like carbon emissions, holding companies and governments accountable for how well they distribute wealth, power, and opportunity.

A Day in This Future

You wake up to check the morning financial reports, but instead of stock indices, the world's major headlines focus on Resilience Index scores—a new global metric that dictates how inclusive and equitable organizations, cities, and countries really are. Businesses, governments, and institutions are now ranked not just by their financial performance but by how well they distribute opportunity, wealth, and decision-making power.

A young entrepreneur in Brazil receives an unexpected surge in funding after her startup ranks in the top 1% of companies for fair wage distribution and accessibility. Meanwhile, a multinational corporation faces public backlash after its Resilience Index score drops due to exploitative supply chain practices—leading investors to pull out and employees to leave for better-ranked competitors. Governments tie public procurement contracts to inclusivity scores, ensuring that infrastructure, healthcare, and urban projects benefit marginalized communities first.

What started as a social accountability movement in response to global inequalities turned into a systemic shift in how success is measured. Much like carbon emissions, companies that fail to meet minimum inclusivity thresholds are fined, whereas those that exceed benchmarks are rewarded with tax incentives and public goodwill. The world is no longer impressed by innovation alone—it demands that progress be equitable.

Switches That Enable This Future

- A major climate catastrophe forces governments to rethink economic inequalities, tying resilience to inclusion metrics.

- A wave of financial instability proves that companies with equitable governance structures are more resilient in economic downturns.

- A breakthrough in real-time social impact tracking technology enables precise measurement of corporate inclusivity efforts.

Critical Question

Can inclusion be effectively quantified, or will these metrics become another tool for performative compliance?

How Leaders, Decision-Makers, and Designers Should Use These Scenarios

Scenarios are not just intellectual exercises; they are powerful tools for reshaping decision-making, influencing strategy, and sparking critical conversations about the future. By presenting possible trajectories, they help leaders and organizations step outside their immediate concerns and imagine the long-term implications of their choices today. This is particularly crucial in an era where social, economic, and technological changes occur at an unprecedented pace, often disrupting traditional business models, governance structures, and cultural norms.

For designers, thinking about the future is not a new concept. The very essence of design involves envisioning possibilities, anticipating needs, and constructing experiences that do not yet exist. Designers routinely engage in speculative work, whether it's prototyping a product that will enter the market years later or shaping urban spaces meant to serve generations to come. The practice of Design Fiction and Future Modeling extends this logic, using narrative, speculative artifacts, and scenario-building to probe deeper societal questions.

I have witnessed firsthand how this approach can challenge deeply entrenched worldviews and lead to significant shifts in strategy. One of the most striking examples comes from my work with one of the largest luxury companies in the world, where I applied my Future Modeling framework. The exercise began as a speculative exploration of what luxury might look like in a world without economic inequality. At first, the company's executives embraced the thought experiment lightly, seeing it as an intellectual challenge rather than a potential disruption to their industry. However, as we began to construct a world where economic disparities had disappeared, an uncomfortable realization took hold: luxury as they knew it barely existed in such a future. Without stark contrasts in wealth, without the aspirational gap that made luxury desirable, many of their most valuable products lost relevance.

The implications of this realization were profound. If luxury was so deeply tied to inequality, then the industry was not just a passive observer but an active participant in maintaining and benefiting from social stratification. This prompted an intense discussion within the company. Rather than seeing this insight as a threat, it began to redefine its role in society. The company explored new ways of framing luxury—not as an exclusive commodity accessible to the few but as an expression of craftsmanship, joy, tradition, and artistic excellence that could be more widely appreciated. Over time, this shift led the firm to reimagine its events, communications, and even product lines, making space for broader participation while preserving its legacy of excellence. The company hosted events that were open to a much wider audience, celebrated the artistry and heritage behind its products, and started telling stories about the beauty of craft rather than exclusivity.

However, opening events to a wider audience is not, in itself, a sufficient solution to inequality. Although it was a step toward greater awareness and inclusion, it did not fundamentally address the systemic disparities that had shaped the industry. The company recognized that inclusivity could not just be about optics: it had to be about real, systemic change. This led to further internal debates and policy shifts focusing on how value is distributed within the industry, how artisans and workers across the supply chain are recognized and compensated, and how luxury could become a vehicle for cultural exchange and preservation rather than a symbol of privilege. The process was neither instant nor without resistance, but the exercise fundamentally reshaped how leadership viewed their industry and its responsibility to society.

Another experience that reinforced the power of Design Fiction came from my work with the French government, specifically with the team overseeing the transformation of public forces and services. Here, we used Future Modeling to explore the evolution of disability and what accessibility might mean in the coming decades. One of the fundamental breakthroughs came when we considered a future in which disability no longer carried its historical stigma but was instead seen as a source of enhanced capabilities. Advances in bionics, exoskeletons, and neurotechnology introduced a possibility where what we today consider disabilities—such as limited mobility or the loss of a limb—could be transformed into functional advantages, even surpassing "normal" human abilities.

This realization led to an entirely new way of thinking about the role of disabled individuals in society. If disabilities could be augmented into hyper-abilities, then the fundamental framework that defined disability in legal and social terms would need to be rewritten. At the same time, we recognized that new forms of disabilities would emerge, shaped by lifestyle diseases, neurodivergence, and evolving mental health challenges. The conversation expanded beyond physical impairments to include the growing prevalence of autism, obesity-related mobility issues, and neurocognitive conditions. Suddenly, the rigid definitions that had governed accessibility laws and urban planning for decades felt outdated and insufficient for the future that was emerging.

One of the most immediate and tangible outcomes of this work was a push to redesign disability signage in public spaces. At the time, the universal disability symbol—a figure in a wheelchair—did not reflect the vast majority of disabled individuals. Moreover, the official designation on many public signs, "GIG-GIC" (reserved for those gravely injured in war or civil life), was a relic of a past era when physical disabilities were largely defined by war wounds. It became clear that this terminology no longer reflected the reality of modern disabilities. Our work led to policy recommendations aimed at updating these symbols and terminology, ensuring that they aligned with the expanding spectrum of disabilities and making public spaces more truly inclusive. Beyond signage, this project spurred further research and governmental initiatives to address the nuances of hidden disabilities—from sensory disorders to chronic pain conditions—ensuring that the future of accessibility was shaped with forward-thinking policies rather than outdated assumptions.

Both of these experiences demonstrated that scenarios are not about predicting the future: they are about expanding the range of what we consider possible. By immersing decision-makers in unfamiliar but plausible realities, they encourage deep reflection, critical examination, and, ultimately, more thoughtful strategies. Organizations that embrace this approach gain a significant advantage: they are not just reacting to change but actively shaping the world they want to exist in.

For businesses, governments, and institutions, the value of speculative futures lies in their ability to break free from linear thinking and short-termism. It forces leaders to engage with ideas that may seem radical, improbable, or even uncomfortable today but could very well define the next era of human society. By integrating these practices into strategic planning, organizations equip themselves with the foresight necessary to navigate uncertainty, make responsible choices, and ensure that their innovations serve a future that is equitable, inclusive, and just.

In the end, the question is not whether the future will change—it always does. The real challenge is whether we are willing to imagine, question, and shape it before it arrives.

Transitioning to Measuring Inclusion—Lessons from Sustainability

Thinking about the future of equity and inclusion inevitably leads to this question: how do we measure progress? In the world of sustainability, we have developed clear metrics—most notably the carbon footprint—to track and assess environmental impact. This standardized measure has provided governments, businesses, and individuals with a tangible way to evaluate their role in climate change. It has structured regulations, guided investments, and created a common language for progress.

But when it comes to diversity, equity, and inclusion (DEI), we lack an equivalent universal metric. Instead, we rely on fragmented measures—workforce demographics, pay equity audits, representation quotas—that, although valuable, fail to capture the full complexity of systemic inclusion. Without clear, universally recognized metrics, progress remains difficult to track, and accountability remains weak.

Throughout my career, I have worked closely with organizations developing sustainability strategies, and I have seen how standardization around carbon has both accelerated and constrained progress. The concept of carbon tunnel vision (see Figure 10.1), where sustainability efforts become narrowly focused on carbon emissions at the expense of broader ecological and social issues, is a powerful example of how metrics shape behavior. Companies optimize for what they can measure, often overlooking essential but less quantifiable aspects of sustainability, such as biodiversity loss, regenerative agriculture, and the social impact of climate change on marginalized communities.

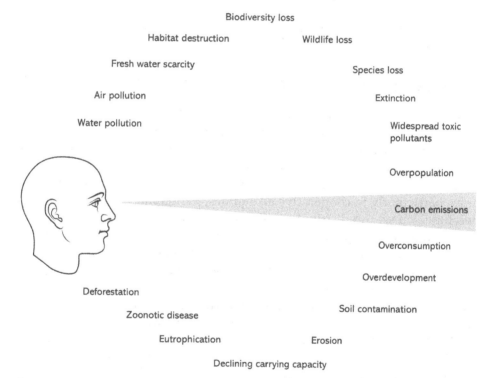

Figure 10.1: Carbon tunnel vision

The same risk applies to DEI metrics. If we focus too much on simplistic diversity numbers—such as percentages of underrepresented groups in the workforce—we may miss deeper structural inequalities. Just as carbon metrics led companies to invest in carbon offsets rather than transforming their business models, DEI efforts can become box-checking exercises instead of driving real change.

To move beyond this limitation, we need more systemic ways to measure inclusion, which is where the concepts of Distributivity and Additionality come into play. Rather than focusing solely on who is represented, these metrics assess how value is distributed and whether new initiatives create positive change rather than merely shifting resources around.

In the next section, we will explore how Distributivity and Additionality can serve as new standards for measuring inclusion, helping us design solutions that go beyond optics and truly reshape economic and social structures for a more equitable future.

One of the most compelling projects that can illustrate the concept of Distributivity is the initiative around Digital Inclusion for Unbanked Citizens from the Equitable Design deck. Although financial inclusion is often discussed in terms of economic opportunities and access to banking services, applying the lens of Distributivity allows us to assess who truly benefits from these efforts and whether they are structured to empower communities rather than merely expand corporate reach.

Distributivity: Measuring How Value Is Shared

Distributivity is the measure of how value—whether financial, social, or infrastructural—is shared across different stakeholders within a system. It asks whether the benefits of an initiative or innovation are concentrated in the hands of a few or fairly dispersed to create broader, systemic equity.

The idea of Distributivity has its roots in economic and social justice theories, particularly in cooperative economics, participatory governance, and decentralized wealth distribution. Thinkers such as Amartya Sen and Elinor Ostrom have long explored models of economic justice where resources and power are shared rather than hoarded at the top. Cooperatives, commons-based peer production, and participatory budgeting are all real-world applications of distributive economic models.

Amartya Sen's work on distributive justice centers around his capability approach, which emphasizes the importance of individual freedoms and opportunities rather than just income or resources.[1] Sen argues that development should be seen as a process of expanding the real freedoms that people enjoy rather than merely increasing economic growth or GDP.[2]

Sen's concept of "capabilities" refers to the actual ability of individuals to achieve various valuable functions as a part of living. This approach shifts the focus from mere possession of resources to the freedom to pursue one's goals. By emphasizing capabilities, Sen's framework allows for a more nuanced understanding of poverty and inequality, considering factors such as education, health, and political participation alongside economic measures.

In his book *Development as Freedom* (Oxford University Press, 2001), Sen argues that freedom is both the primary end and the principal means of development.

He contends that expanding individual freedoms should be the main goal of development efforts and the chief way to achieve it.

Elinor Ostrom's work focuses on common-pool resources and collective action, challenging the notion that shared resources inevitably lead to overexploitation (the "tragedy of the commons").[3] Her research demonstrates that communities can effectively manage shared resources through self-governing institutions without resorting to privatization or state control.

Ostrom's book *Governing the Commons* (Cambridge University Press, 1990) presents numerous case studies of successful community-managed resources, from irrigation systems to fisheries. She identifies design principles for effective management of common-pool resources, emphasizing the importance of clear boundaries, collective-choice arrangements, and nested enterprises.[4]

Ostrom's work on polycentric governance systems further explores how multiple decision-making centers at different scales can contribute to more effective and equitable resource management. This approach aligns with the concept of Distributivity by promoting decentralized decision-making and power-sharing.

The concept of Distributivity extends this thinking into the realm of design, business models, and innovation, asking "How do we ensure that the value generated by new ideas, products, or policies is widely shared rather than concentrated among a privileged few?" This is particularly relevant in an era where economic and technological gains have increasingly accumulated at the top, exacerbating inequality across industries and regions.

Although these ideas have been explored in various disciplines, my approach to Distributivity has emerged through years of working in innovation, design, and equitable business strategies. The framework described next reflects how I have applied and refined the concept in real-world projects to measure and drive systemic inclusivity.

How to Measure Distributivity

To quantify Distributivity, we must look beyond surface-level diversity metrics and focus on deeper structural indicators. Some key questions that help assess Distributivity include the following:

- **Who benefits from the initiative?** What proportion of the stakeholders experience positive change?

- **How is value shared?** What percentage of wealth, data access, or decision-making power is concentrated at the top versus redistributed?

- **Is power decentralized?** Are decision-making processes designed to include diverse voices, or are they controlled by a select few?

- **What depth of impact does this initiative have?** Does it create meaningful, long-term change for underrepresented communities, or does it only offer short-term solutions?

Applying Distributivity to the Digital Inclusion for Unbanked Citizens Initiative

In a project mentioned earlier in the book, a major European bank partnered with me and my team to develop financial products aimed at unbanked citizens: those without access to traditional banking services due to economic barriers, lack of documentation, or systemic exclusion. The goal was to create a range of services, including a special debit card, SMS-based banking solutions, and non-digital financial services, ensuring accessibility for individuals without smartphones or stable Internet access.

However, merely providing banking products does not automatically ensure a fairer distribution of financial benefits. This is where the principle of Distributivity comes into play. A distributive approach asks key questions:

- **Who captures the most value in this system?** Is the bank ultimately extracting more from the newly banked customers (through fees, data collection, or reliance on specific banking products), or are these customers genuinely experiencing a shift in financial agency?

- **How widely is the value dispersed?** Are these solutions reaching only a select group of people, or are they accessible to diverse populations with different needs?

- **Are there mechanisms to counterbalance power concentration?** Does this initiative empower local community financial networks, credit unions, or social organizations to have a stake in this financial inclusion effort, or does it reinforce corporate monopolies?

Expanding the Impact Through Distributivity

A truly distributive model would involve:

- **Community-owned banking models:** Instead of just expanding the bank's customer base, could there be cooperative banking models that allow local communities to hold a stake in financial services, redistributing value more equitably?

- **Reinvestment of fees into local economies:** Could transaction fees be reinvested into microloans or local development projects, ensuring that financial flows support grassroots economic growth rather than purely corporate gains?

- **Education and agency building:** Financial literacy programs, co-created with communities, could be integrated into the service, ensuring that individuals are not just passive users but active agents in managing their financial futures.

- **Partnerships with alternative banking structures:** Working with credit unions, microfinance initiatives, or cooperatives that distribute decision-making power rather than centralizing it.

By applying Distributivity as a guiding framework, digital financial inclusion can shift from being a corporate expansion strategy to a system that genuinely redistributes financial power to those who need it most. This model can then serve as a replicable blueprint for other industries seeking to create equitable and sustainable inclusion initiatives.

Measuring Distributivity: Practical Approaches

Although Distributivity is a powerful concept, it must be measurable to be actionable. Without clear indicators, it risks becoming an abstract ideal rather than a tangible framework for decision-making. Following are accessible ways to quantify Distributivity, ensuring that it can be systematically applied in business models, policy initiatives, and design processes.

Some of these tools draw from widely recognized calculation methodologies, others are adaptations or refinements, and some are original frameworks I have developed.

To simplify, we can use three primary indicators to assess how value is shared within a system:

- **The number of people benefiting from an initiative:** A distributive model should impact a broad range of stakeholders, not just a privileged few. Measuring how many people gain real, tangible benefits from a policy, innovation, or investment gives a basic but powerful indication of Distributivity.

- **The share of value captured by different groups:** This can be analyzed using income or resource distribution data. A useful tool here is the Gini coefficient, a widely recognized measure of inequality. The Gini coefficient, also known as the Gini index or Gini ratio, is a widely used statistical measure of economic inequality within a population. A great source for understanding the Gini coefficient is the book *Variabilità e mutabilità* (*Variability and Mutability*), published in 1912 by Italian statistician Corrado Gini, who developed this measure.[5] A lower Gini coefficient suggests that value is more evenly distributed, whereas a higher coefficient indicates concentration of wealth and power.

- **Decentralization of power and decision-making:** A truly distributive initiative does not centralize control in a single entity; instead, multiple groups have a say in shaping decisions. This can be measured through participation levels, voting rights, or governance structures.

A Simple Formula for Distributivity

One way to quantify Distributivity in a business or social enterprise is to measure how much value is shared versus retained at the top:

$$\text{Distributivity Index} = (\text{Revenue Shared with Employees} + \text{Community Reinvestment}) / \text{Total Revenue}$$

For example, if 80% of revenue is absorbed by executives and investors, whereas only 20% goes to employees, local businesses, and community programs, the Distributivity index is low.

Conversely, if 60% of revenue is reinvested in wages, community infrastructure, and cooperative ownership models, the index is high, signaling better Distributivity.

Applying Distributivity to Real-World Scenarios

Let's take the preceding digital financial inclusion project as an example. This initiative sought to provide financial tools for unbanked citizens, but applying a Distributivity lens raises deeper questions:

- How many people actually gained access, and who remained excluded?
- Did the financial institution retain most of the profit, or were transaction fees reinvested in local economies?
- Were communities involved in governance, or was decision-making controlled by corporate interests?

A truly distributive model in this case would not only provide financial access but also ensure fair profit-sharing, reinvestment in communities, and local ownership of financial decision-making.

Why This Matters

Measuring Distributivity is essential because economic and social systems often default to concentrating power rather than sharing it. By embedding these measurements into project evaluations, policy frameworks, and corporate governance, we ensure that Distributivity is not just an aspiration but a practical, measurable goal.

Advanced Methods for Measuring Distributivity

Although the Distributivity Index provides a simple way to evaluate how value is shared, other quantitative methods offer deeper insights into whether an initiative or system is truly distributive.

The Distributive Ratio (DR)

One way to measure Distributivity in a system is to compare the spread of benefits among different stakeholders. This can be captured through a Distributive Ratio (DR):

$$DR = (\text{Sum of Benefits Received by All Stakeholders}) /$$
$$(\text{Total Number of Stakeholders})$$

where:

a higher DR suggests a more equitable spread of benefits, whereas a lower DR indicates that value is concentrated in the hands of a few.

For example, if a corporate social responsibility (CSR) program donates resources to 10,000 beneficiaries, but 90% of funds go to just 10% of the beneficiaries, the DR will be low, meaning that despite large participation, the program is failing in its Distributivity goals.

The Lorenz Curve and Gini Coefficient

A more statistically rigorous approach is to use the Lorenz curve and the Gini coefficient, which are commonly applied to measure income inequality but highly useful for evaluating Distributivity in social impact projects, technology, and corporate governance.

The Gini coefficient (G) is defined as:

$$G = 1 - 2 \times (\text{integral of Lorenz curve})$$

where:

- The Lorenz curve, shows the cumulative share of total value owned by a given percentage of the population.
- $G = 0 \rightarrow$ Perfect Distributivity (everyone has an equal share).
- $G = 1 \rightarrow$ Extreme concentration (one entity holds all the value).

Here is an example in product design:

- If one large retailer controls 80% of all small supplier contracts, the Gini coefficient will be close to 1, showing extreme concentration.
- If contracts are spread evenly across 100 suppliers, the coefficient will be lower, indicating a more distributive system.

The Decentralization Decision Index

To measure decision-making power, we can use a Decentralization Index (DI) to assess whether power is concentrated at the top or distributed among multiple groups:[6]

DI = (Number of Decisions Made at Local Levels) /

(Total Number of Decisions in the System)

- A higher DI indicates that power is widely distributed, meaning multiple stakeholders have a say in governance.
- A low DI suggests centralized control, where only a few people or entities make key decisions.

Here's an example in policy design:

- A public budget allocation system where 50% of funds are distributed via community votes will have a high DI.
- A company where only executives set diversity policies without consulting employees will have a low DI, indicating poor Distributivity in decision-making.

The Equity-of-Access Index (EAI)

This measure is useful for evaluating projects that aim to increase access to essential services, such as education, healthcare, or financial inclusion:

EAI = (Number of Previously Excluded Individuals Who Gained Access) /

(Total Target Population)

- Higher EAI values indicate that an initiative is genuinely reducing barriers to inclusion.
- Lower EAI values suggest that the effort benefits only those who were already somewhat included.

Here's an example in digital access:

- If a telecom company launches a new Internet initiative for rural areas, but only 10% of the unconnected population gains access, the EAI will be low, suggesting that the project is not truly distributive.
- If 70% of the previously unconnected population gains access, the EAI is high, showing real progress toward equitable distribution.

Applying These Measures in Real-World Contexts

Let's take an example from my body of work presented earlier: a project providing employment pathways for marginalized communities.

Using the different Distributivity measures:

- **Distributive Ratio (DR):** Measures how many people actually experience career mobility versus how many remain in low-wage positions
- **Gini Coefficient (G):** Analyzes whether wages and promotions are equitably spread or concentrated in higher-level roles
- **Decentralization Index (DI):** Evaluates whether employees have input into workplace policies or if leadership makes unilateral decisions
- **Equity-of-Access Index (EAI):** Tracks how many previously excluded workers gain real access to professional development

Each of these quantitative tools ensures that Distributivity is not just an abstract goal but a measurable reality that can drive better policy, design, and corporate responsibility.

Additionality: Ensuring True Impact Rather than Displacement

Additionality is a critical measure of whether an initiative, innovation, or policy creates new value rather than simply replacing or displacing existing efforts. It assesses whether an intervention brings something genuinely needed and beneficial to a system or merely shifts resources without solving underlying challenges.

The concept of Additionality has its roots in economics, public policy, and environmental finance, particularly in carbon offset programs and social impact assessments. Scholars and institutions, such as Nicholas Stern (author of *The Economics of Climate Change,* Cambridge University Press, 2007) and the Organization for Economic Cooperation and Development (OECD), have long explored Additionality in the context of public investment, subsidies, and sustainability initiatives. In environmental finance, for example, carbon offset projects must demonstrate that their impact would not have occurred without external funding—ensuring true Additionality.

In a business or policy context, Additionality ensures that interventions do not cannibalize existing positive efforts or reinforce systemic inequities. It requires us to critically examine whether an initiative:

- Introduces new solutions that were previously unavailable
- Improves access to essential resources without eliminating community-led efforts
- Enhances systemic equity rather than reinforcing existing power structures

Although these ideas have been explored across various fields, my own approach to Additionality has evolved through years of working in innovation, design, and equitable business strategies. The framework that follows reflects how I have applied Additionality in real-world projects to ensure that new initiatives truly add value.

How to Measure Additionality

To quantify Additionality, we must move beyond vanity metrics and assess whether an initiative fills a genuine gap, enhances systemic fairness, and avoids displacing existing beneficial structures.

Key questions to assess Additionality include the following:

- Was this solution already available in some form? If so, is the new initiative expanding access or replacing an existing effort?
- Does this initiative support or suppress local/community-driven solutions? Is it strengthening grassroots efforts or merely outcompeting them?
- Does the initiative reinforce existing inequities? Could it inadvertently increase dependence on corporate control or lead to a net loss in autonomy?
- What unintended consequences could arise? Are there hidden risks of shifting value away from underserved communities?

Applying Additionality to a Real-World Example

A prime illustration of Additionality is the Nike Pro Hijab initiative, which aimed to provide sports-friendly hijabs for Muslim athletes. Although the intention was noble, the initiative displaced small, independent businesses that had long been designing culturally specific, community-driven sportswear.

Applying an Additionality lens, we ask:

- Did this innovation truly fill a gap, or did it commercialize an existing solution?
- Could Nike have partnered with smaller, community-driven brands rather than replacing them?
- Did the initiative increase access, or did it consolidate power further in a global corporation?

A more additional approach would have involved:

- Collaborating with community-led brands to enhance their reach
- Reinvesting profits into grassroots Muslim sports organizations
- Ensuring that traditional, small-scale producers were not erased by corporate dominance

By applying Additionality, we avoid extractive innovation models and create systems where new initiatives complement rather than displace existing solutions.

Measuring Additionality: Practical Approaches

To move beyond theory, Additionality must be measurable and actionable. The following are accessible ways to quantify it.

Additionality Score (AS)

A basic way to measure Additionality is through an Additionality Score, which compares the net benefit of an initiative to the extent of preexisting solutions:

$$AS = (\text{New Value Created} - \text{Preexisting Value Displaced}) /$$
$$(\text{Total Impact of Initiative})$$

- A high AS indicates that the initiative is truly adding value.
- A low AS suggests that it may be replacing or shifting rather than creating.

Here's an example:

- If a tech startup creates an app for farmers but replaces community-led knowledge-sharing networks, the AS is low.
- If the app enhances these networks by digitizing them, ensuring collaborative growth, the AS is high.

Counterfactual Analysis

One of the most rigorous methods for measuring Additionality is a counterfactual analysis,[7] which asks "What would have happened if this initiative did not exist?"

- If similar benefits would have occurred anyway, the initiative lacks Additionality.
- If the initiative creates a completely new and necessary solution, it demonstrates high Additionality.

Here's an example in urban planning:

- If a large real estate project claims to provide affordable housing, but the same level of housing development would have occurred without it, then the Additionality is questionable.
- If the project builds housing that would not have been developed otherwise, the Additionality is high.

Displacement Index (DI)

To measure whether an initiative displaces existing efforts, we use a Displacement Index (DI):

$$DI = (\text{Preexisting Initiatives Displaced}) / (\text{Total New Initiative Impact})$$

- A high DI signals that the initiative is replacing rather than adding.
- A low DI means the initiative is creating a net new impact.

Here's an example from social programs:

- If an international NGO launches a food security initiative but inadvertently outcompetes local farmers and food programs, the DI is high—indicating low Additionality.
- If it builds local capacity and strengthens community-led programs, the DI is low, signaling true Additionality.

Reinforcement Index (RI)

This measures whether an initiative strengthens existing structures rather than replacing them:

$$RI = (\text{Value Reinforced for Existing Solutions}) / (\text{Total New Value Created})$$

- A high RI means the initiative supports and amplifies current efforts.
- A low RI suggests that the initiative is undermining community-based models.

Here's an example in healthcare innovation:

- If a corporate health-tech solution replaces traditional clinics, the RI is low, meaning it fails to enhance local structures.
- If it integrates into existing systems, making services more efficient, the RI is high.

Applying These Measures in Real-World Contexts

Let's take an example mentioned previously: an initiative focused on bringing digital education to underprivileged communities. Using Additionality measures:

- **Additionality Score (AS):** Measures whether the project introduces net new educational opportunities or just digitizes existing teaching methods

- **Counterfactual Analysis:** Examines whether education access would have improved without this initiative

- **Displacement Index (DI):** Determines whether the new initiative erases local educational programs

- **Reinforcement Index (RI):** Evaluates whether the initiative is strengthening local educators or replacing them with external digital content

By embedding these quantitative assessments, we ensure that Additionality is not just a buzzword but a practical framework for assessing impact.

Why Additionality Matters

In a world where corporations, governments, and institutions constantly introduce new initiatives, Additionality helps us differentiate between true progress and mere displacement. By measuring Additionality, we ensure that:

- New initiatives create value rather than extract it.

- Communities retain agency over their solutions.

- Innovations complement rather than replace grassroots efforts.

In the next section, we will explore how Distributivity and Additionality work together, forming a dual metric for equitable innovation that ensures the fair sharing of value (Distributivity) and the genuine creation of new opportunities (Additionality).

The Dual Framework: Using Distributivity and Additionality Together

For an initiative to create real, systemic change, it must both distribute value equitably and introduce new value rather than just shift resources around. In other words, Distributivity and Additionality must work together. Measuring only one without considering the other can lead to incomplete or harmful interventions.

Why One Metric Alone Is Not Enough

If an initiative scores high on Distributivity but low on Additionality, it may spread benefits widely but fail to create new opportunities. This can result in redistribution without transformation—an approach that may feel impactful in the short term but does not challenge deeper systemic inequalities.

Conversely, if an initiative is high on Additionality but low on Distributivity, it may introduce valuable new solutions, but these benefits are concentrated in the hands of a few, exacerbating existing inequities rather than alleviating them.

A truly inclusive and Equitable Design approach must strive for both high Distributivity and high Additionality—ensuring that new initiatives not only create impact but also share that impact broadly and fairly.

The Conceptual Matrix

One way to visualize the relationship between Distributivity and Additionality is through a matrix that categorizes initiatives based on their balance of these two metrics:

	LOW ADDITIONALITY (NO NEW VALUE CREATED)	HIGH ADDITIONALITY (NEW VALUE CREATED)
Low Distributivity (value is concentrated)	Exploitative systems (e.g., monopolistic AI systems that amplify biases rather than reducing them)	Corporate-driven disruption (e.g., cutting-edge tech that benefits only the wealthy few)
High Distributivity (value is shared widely)	Redistribution without transformation (e.g., charity programs that provide aid but don't change systemic conditions)	Ideal transformation (e.g., equitable policies, cooperatively designed tech that creates new opportunities and shares them fairly)

Real-World Applications

Organizations, policymakers, and designers must evaluate their initiatives across both metrics. Some examples include the following:

- **High Distributivity, low Additionality:** A universal basic income program that redistributes wealth but does not create new economic opportunities
- **Low Distributivity, high Additionality:** A groundbreaking AI tool that automates medical diagnostics but remains accessible only to elite institutions
- **High Distributivity, high Additionality:** A cooperative-owned fintech platform that expands access to financial services while redistributing profits to communities

By embedding Distributivity and Additionality into impact assessments, organizations can move beyond surface-level inclusivity toward truly transformative and equitable solutions.

A Future-Proof Approach to Inclusive Design

We cannot predict the future, but we can prepare for it. History has shown that unexpected crises, technological shifts, and social movements can rapidly redefine societal priorities. The next major shift—whether political, technological, or environmental—may challenge existing systems in ways we cannot yet foresee. Although we cannot anticipate every possible challenge, we can adopt a mindset that prepares us for the unknown by building resilience through Inclusive Design.

This is why investing in systemic, long-term inclusivity is the most reliable strategy for future-proofing organizations, communities, and societies. Unlike reactive approaches, which respond only after crises occur, proactive design ensures resilience and adaptability in an unpredictable world. Inclusive Design is not merely a short-term response to diversity concerns: it is an investment in long-term stability, adaptability, and fairness. When inclusivity is woven into the DNA of decision-making processes, institutions become more equipped to handle societal shifts without resorting to reactive, temporary fixes.

As we have seen with global crises such as pandemics, economic recessions, and climate disasters, these events disproportionately affect marginalized populations. When systems are designed without considering Distributivity and Additionality, they fail the very people who need them most. Economic downturns push vulnerable workers further into precarity, climate-related disasters displace communities with the least resources to recover, and technological advancements widen existing inequalities when accessibility is not prioritized. Organizations and governments that recognize these patterns have an opportunity to preempt future disparities by embedding inclusivity into their strategies today.

Why Inclusive Design Is the Best Future Strategy

- Shocks and disruptions (e.g., global pandemics, climate migration, economic crises) tend to amplify inequalities. Systems that are already inclusive and distributive can better withstand these shocks.

- Emerging technologies (e.g., AI, biotechnology) have the potential to either broaden or restrict access. If inclusivity is not built in from the start, these tools risk exacerbating inequities.

- Social movements consistently push for greater equity, whether in labor rights, disability advocacy, or racial justice. Organizations that resist these changes face long-term instability and reputational damage.

The Role of Governments, Businesses, and Communities

To make Inclusive Design systemic, all sectors must play a role:

- Governments must implement policy frameworks that embed Distributivity and Additionality into regulations, ensuring that public and private initiatives serve broad social good rather than reinforcing existing inequalities.

- Businesses must move beyond performative DEI efforts and integrate Distributivity and Additionality into their core operations, not just as compliance measures but as strategic imperatives.

- Communities must be empowered to co-create solutions, ensuring that innovations reflect real, lived experiences rather than being imposed from the top down.

Inclusive Design is not just about rectifying past inequities: it is about building a world that is fundamentally resilient to future challenges. When we prioritize Distributivity, we ensure that the benefits of progress are widely shared. When we prioritize Additionality, we guarantee that innovations create true impact rather than merely shifting existing resources. The convergence of these two principles forms the foundation of an equitable, future-ready society.

A Call to Action: Moving from Reactive to Proactive Approaches

The future of Inclusive Design is not just about solving today's problems—it is about ensuring that our systems, businesses, and policies remain resilient and just in the face of tomorrow's challenges.

By combining the principles of Distributivity and Additionality, decision-makers, designers, and policymakers can move beyond incremental fixes and toward a more just, equitable, and future-ready society.

This requires a fundamental shift in how we approach innovation, design, and economic structures. Rather than waiting for crises to force changes, we must actively design for inclusion, fairness, and equity from the outset. The work of embedding Distributivity and Additionality into institutions must begin now, ensuring that inclusivity is not an afterthought but an essential feature of our world moving forward.

Conclusion: Designing a Just and Equitable Future

As this book has explored, Inclusive Design is not simply about making spaces, products, and services accessible—it is about fundamentally reshaping our systems to work for everyone. Achieving true equity requires us to rethink who holds power, who benefits from innovation, and how resources are distributed. It is a challenge that demands bold thinking, collective effort, and sustained commitment.

The concepts of Distributivity and Additionality provide a lens through which we can assess and shape our work moving forward. They remind us that true progress is measured not only by what we create but also by how widely and fairly it is shared. The most successful organizations, policies, and innovations of the future will be those that embrace these principles, creating solutions that uplift communities rather than extract from them.

But the responsibility does not lie with businesses and policymakers alone. Each of us, in our respective roles, can contribute to shaping a more equitable world. Whether through advocacy, design, entrepreneurship, or governance, we all have the capacity to challenge outdated models and push for systemic change.

The future remains unwritten. The paths we take today will define the world we leave behind. Let us choose wisely, designing not just for profit or efficiency but for equity, dignity, and justice—a future where no one is left behind.

Notes

[1] Mboula Wamok-Peter, F. S., 2020. *Amartya Sen et l'Éthique du développement*. PhD thesis. Université de Lille. [Online PDF]. Available at: `https://theses.hal.science/tel-03143063v1/document`.

[2] Wells, T., n.d. Sen's capability approach. Internet Encyclopedia of Philosophy. [Online]. Available at: `https://iep.utm.edu/sen-cap`.

[3] Rayamajhee, V. & Paniagua, P., 2020. Elinor Ostrom and the contestable nature of goods. Centre for the Study of Governance & Society. [Online]. Available at: `https://csgs.kcl.ac.uk/elinor-ostrom-and-the-contestable-nature-of-goods`.

[4] Ostrom, E., 2009. Beyond markets and states: polycentric governance of complex economic systems. Nobel Prize lecture. [Online PDF]. Available at: `https://www.nobelprize.org/uploads/2018/06/ostrom_lecture.pdf`.

[5] World Bank, n.d. Gini index. [Online]. Available at: `https://databank.worldbank.org/metadataglossary/world-development-indicators/series/SI.POV.GINI`.

[6] Saito, Y., 2024. How to measure decentralization: a summary of the global literature review on decentralization measurement. World Bank Group. [Online]. Available at: `https://documents.worldbank.org/en/publication/documents-reports/documentdetail/099041824234020578`.

[7] Le Maux, B., n.d. Counterfactual impact evaluation. [Online]. Available at: `https://sites.google.com/site/benoitlemaux/cours-lectures/counterfactual-impact-evaluation`.

Index